Sludge Treatment and Disposal

Biological Wastewater Treatment Series

The *Biological Wastewater Treatment* series is based on the book *Biological Wastewater Treatment in Warm Climate Regions* and on a highly acclaimed set of best selling textbooks. This international version is comprised by six textbooks giving a state-of-the-art presentation of the science and technology of biological wastewater treatment.

Titles in the *Biological Wastewater Treatment* series are:

Volume 1: *Wastewater Characteristics, Treatment and Disposal*
Volume 2: *Basic Principles of Wastewater Treatment*
Volume 3: *Waste Stabilisation Ponds*
Volume 4: *Anaerobic Reactors*
Volume 5: *Activated Sludge and Aerobic Biofilm Reactors*
Volume 6: *Sludge Treatment and Disposal*

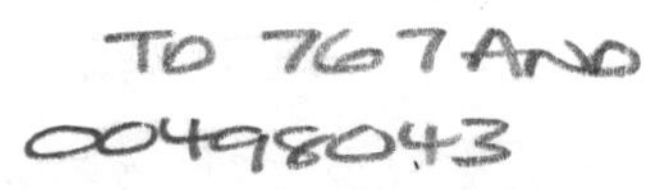

Biological Wastewater Treatment Series

VOLUME SIX

Sludge Treatment and Disposal

Cleverson Vitorio Andreoli, Marcos von Sperling and Fernando Fernandes (Editors)

Publishing
LONDON • NEW YORK

Published by IWA Publishing, Alliance House, 12 Caxton Street, London SW1H 0QS, UK

Telephone: +44 (0) 20 7654 5500; Fax: +44 (0) 20 7654 5555; Email: publications@iwap.co.uk
Website: **www.iwapublishing.com**

First published 2007

Copy-edited and typeset by Aptara Inc., New Delhi, India
Printed by Lightning Source

Disclaimer
The information provided and the opinions given in this publication are not necessarily those of IWA or of the editors, and should not be acted upon without independent consideration and professional advice. IWA and the editors will not accept responsibility for any loss or damage suffered by any person acting or refraining from acting upon any material contained in this publication.

British Library Cataloguing in Publication Data
A CIP catalogue record for this book is available from the British Library

Library of Congress Cataloguing-in-Publication Data
A catalogue record for this book is available from the Library of Congress

ISBN: 1 84339 166 X
ISBN 13: 9781843391661

Contents

Preface

The present series of books has been produced based on the book "*Biological wastewater treatment in warm climate regions*", written by the same authors and also published by IWA Publishing. The main idea behind this series is the subdivision of the original book into smaller books, which could be more easily purchased and used.

The implementation of wastewater treatment plants has been so far a challenge for most countries. Economical resources, political will, institutional strength and cultural background are important elements defining the trajectory of pollution control in many countries. Technological aspects are sometimes mentioned as being one of the reasons hindering further developments. However, as shown in this series of books, the vast array of available processes for the treatment of wastewater should be seen as an incentive, allowing the selection of the most appropriate solution in technical and economical terms for each community or catchment area. For almost all combinations of requirements in terms of effluent quality, land availability, construction and running costs, mechanisation level and operational simplicity there will be one or more suitable treatment processes.

Biological wastewater treatment is very much influenced by climate. Temperature plays a decisive role in some treatment processes, especially the natural-based and non-mechanised ones. Warm temperatures decrease land requirements, enhance conversion processes, increase removal efficiencies and make the utilisation of some treatment processes feasible. Some treatment processes, such as anaerobic reactors, may be utilised for diluted wastewater, such as domestic sewage, only in warm climate areas. Other processes, such as stabilisation ponds, may be applied in lower temperature regions, but occupying much larger areas and being subjected to a decrease in performance during winter. Other processes, such as activated sludge and aerobic biofilm reactors, are less dependent on temperature, as a result of the higher technological input and mechanisation level. The main purpose of this series of books is to present the technologies for urban wastewater treatment as applied to the specific condition of warm temperature, with the related implications in terms of design and operation. There is no strict definition for the range of temperatures that fall into this category, since the books always present how to correct parameters, rates and coefficients for different temperatures. In this sense, subtropical and even temperate climate are also indirectly covered, although most of the focus lies on the tropical climate.

Another important point is that most warm climate regions are situated in developing countries. Therefore, the books cast a special view on the reality of these countries, in which simple, economical and sustainable solutions are strongly demanded. All technologies presented in the books may be applied in developing countries, but of course they imply different requirements in terms of energy, equipment and operational skills. Whenever possible, simple solutions, approaches and technologies are presented and recommended.

Considering the difficulty in covering all different alternatives for wastewater collection, the books concentrate on off-site solutions, implying collection and transportation of the wastewater to treatment plants. No off-site solutions, such as latrines and septic tanks are analysed. Also, stronger focus is given to separate sewerage systems, although the basic concepts are still applicable to combined and mixed systems, especially under dry weather conditions. Furthermore, emphasis is given to urban wastewater, that is, mainly domestic sewage plus some additional small contribution from non-domestic sources, such as industries. Hence, the books are not directed specifically to industrial wastewater treatment, given the specificities of this type of effluent. Another specific view of the books is that they detail biological treatment processes. No physical-chemical wastewater treatment processes are covered, although some physical operations, such as sedimentation and aeration, are dealt with since they are an integral part of some biological treatment processes.

The books' proposal is to present in a balanced way theory and practice of wastewater treatment, so that a conscious selection, design and operation of the wastewater treatment process may be practised. Theory is considered essential for the understanding of the working principles of wastewater treatment. Practice is associated to the direct application of the concepts for conception, design and operation. In order to ensure the practical and didactic view of the series, 371 illustrations, 322 summary tables and 117 examples are included. All major wastewater treatment processes are covered by full and interlinked design examples which are built up throughout the series and the books, from the determination of the wastewater characteristics, the impact of the discharge into rivers and lakes, the design of several wastewater treatment processes and the design of the sludge treatment and disposal units.

The series is comprised by the following books, namely: *(1) Wastewater characteristics, treatment and disposal; (2) Basic principles of wastewater treatment; (3) Waste stabilisation ponds; (4) Anaerobic reactors; (5) Activated sludge and aerobic biofilm reactors; (6) Sludge treatment and disposal.*

Volume 1 (*Wastewater characteristics, treatment and disposal)* presents an integrated view of water quality and wastewater treatment, analysing wastewater characteristics (flow and major constituents), the impact of the discharge into receiving water bodies and a general overview of wastewater treatment and sludge treatment and disposal. Volume 1 is more introductory, and may be used as teaching material for undergraduate courses in Civil Engineering, Environmental Engineering, Environmental Sciences and related courses.

Volume 2 (*Basic principles of wastewater treatment)* is also introductory, but at a higher level of detailing. The core of this book is the unit operations and processes associated with biological wastewater treatment. The major topics covered are: microbiology and ecology of wastewater treatment; reaction kinetics and reactor hydraulics; conversion of organic and inorganic matter; sedimentation; aeration. Volume 2 may be used as part of postgraduate courses in Civil Engineering, Environmental Engineering, Environmental Sciences and related courses, either as part of disciplines on wastewater treatment or unit operations and processes.

Volumes 3 to 5 are the central part of the series, being structured according to the major wastewater treatment processes (*waste stabilisation ponds, anaerobic reactors, activated sludge and aerobic biofilm reactors*). In each volume, all major process technologies and variants are fully covered, including main concepts, working principles, expected removal efficiencies, design criteria, design examples, construction aspects and operational guidelines. Similarly to Volume 2, volumes 3 to 5 can be used in postgraduate courses in Civil Engineering, Environmental Engineering, Environmental Sciences and related courses.

Volume 6 (*Sludge treatment and disposal*) covers in detail sludge characteristics, production, treatment (thickening, dewatering, stabilisation, pathogens removal) and disposal (land application for agricultural purposes, sanitary landfills, landfarming and other methods). Environmental and public health issues are fully described. Possible academic uses for this part are same as those from volumes 3 to 5.

Besides being used as textbooks at academic institutions, it is believed that the series may be an important reference for practising professionals, such as engineers, biologists, chemists and environmental scientists, acting in consulting companies, water authorities and environmental agencies.

The present series is based on a consolidated, integrated and updated version of a series of six books written by the authors in Brazil, covering the topics presented in the current book, with the same concern for didactic approach and balance between theory and practice. The large success of the Brazilian books, used at most graduate and post-graduate courses at Brazilian universities, besides consulting companies and water and environmental agencies, was the driving force for the preparation of this international version.

In this version, the books aim at presenting consolidated technology based on worldwide experience available at the international literature. However, it should be recognised that a significant input comes from the Brazilian experience, considering the background and working practice of all authors. Brazil is a large country with many geographical, climatic, economical, social and cultural contrasts, reflecting well the reality encountered in many countries in the world. Besides, it should be mentioned that Brazil is currently one of the leading countries in the world on the application of anaerobic technology to domestic sewage treatment, and in the post-treatment of anaerobic effluents. Regarding this point, the authors would like to show their recognition for the Brazilian Research Programme on Basic Sanitation (PROSAB), which, through several years of intensive, applied, cooperative research has led to the consolidation of anaerobic treatment and aerobic/anaerobic post-treatment, which are currently widely applied in full-scale plants in Brazil. Consolidated results achieved by PROSAB are included in various parts of the book, representing invaluable and updated information applicable to warm climate regions.

Volumes 1 to 5 were written by the two main authors. Volume 6 counted with the invaluable participation of Cleverson Vitorio Andreoli and Fernando Fernandes, who acted as editors, and of several specialists, who acted as chapter authors: Aderlene Inês de Lara, Deize Dias Lopes, Dione Mari Morita, Eduardo Sabino Pegorini, Hilton Felício dos Santos, Marcelo Antonio Teixeira Pinto, Maurício Luduvice, Ricardo Franci Gonçalves, Sandra Márcia Cesário Pereira da Silva, Vanete Thomaz Soccol.

Many colleagues, students and professionals contributed with useful suggestions, reviews and incentives for the Brazilian books that were the seed for this international version. It would be impossible to list all of them here, but our heartfelt appreciation is acknowledged.

The authors would like to express their recognition for the support provided by the Department of Sanitary and Environmental Engineering at the Federal University of Minas Gerais, Brazil, at which the two authors work. The department provided institutional and financial support for this international version, which is in line with the university's view of expanding and disseminating knowledge to society.

Finally, the authors would like to show their appreciation to IWA Publishing, for their incentive and patience in following the development of this series throughout the years of hard work.

Marcos von Sperling
Carlos Augusto de Lemos Chernicharo

December 2006

The authors

CHAPTER AUTHORS

Aderlene Inês de Lara, PhD. Paraná Water and Sanitation Company (SANEPAR), Brazil. aderlene@funpar.ufpr.br

Cleverson Vitório Andreoli, PhD. Paraná Water and Sanitation Company (SANEPAR), Brazil. c.andreoli@sanepar.com.br

Deize Dias Lopes, PhD. Londrina State University (UEL), Brazil. dilopes@uel.br

Dione Mari Morita, PhD. University of São Paulo (USP), Brazil. dmmorita@usp.br

Eduardo Sabino Pegorini. Paraná Water and Sanitation Company (SANEPAR), Brazil. epegorini@sanepar.com.br

Fernando Fernandes, PhD. Londrina State University (UEL), Brazil. fernando@uel.br

Hílton Felício dos Santos, PhD. Consultant, Brazil. hfsantos@uol.com.br

Marcelo Antonio Teixeira Pinto, MSc. Federal District Water and Sanitation Company (CAESB), Brazil. marcelo.teixeira@mkmbr.com.br

Marcos von Sperling, PhD. Federal University of Minas Gerais, Brazil. marcos@desa.ufmg.br

Maurício Luduvice, PhD. MSc. Federal District Water and Sanitation Company (CAESB), Brazil. luduvice@br.inter.net

Ricardo Franci Gonçalves, PhD. Federal University of Espírito Santo, Brazil. franci@npd.ufes.br

Sandra Márcia Cesário Pereira da Silva, PhD. Londrina State University (UEL), Brazil. sandra@uel.br

Vanete Thomaz Soccol, PhD. Federal University of Paraná (UFPR), Brazil. vasoccol@bio.ufpr.br

1

Introduction to sludge management

M. von Sperling, C.V. Andreoli

The management of sludge originating from wastewater treatment plants is a highly complex and costly activity, which, if poorly accomplished, may jeopardise the environmental and sanitary advantages expected in the treatment systems. The importance of this practice was acknowledged by Agenda 21, which included the theme of environmentally wholesome management of solid wastes and questions related with sewage, and defined the following orientations towards its administration: reduction in production, maximum increase of reuse and recycling, and the promotion of environmentally wholesome treatment and disposal.

The increasing demands from society and environmental agencies towards better environmental quality standards have manifested themselves in public and private sanitation service administrators. Due to the low indices of wastewater treatment prevailing in many developing countries, a future increase in the number of wastewater treatment plants is naturally expected. As a consequence, the amount of sludge produced is also expected to increase. Some environmental agencies in these countries now require the technical definition of the final disposal of sludge in the licensing processes. These aspects show that solids management is an increasing matter of concern in many countries, tending towards a fast-growing aggravation in the next years, as more wastewater treatment plants are implemented.

ISBN: 1 84339 166 X. Published by IWA Publishing, London, UK.

The term '**sludge**' has been used to designate the solid by-products from wastewater treatment. In the biological treatment processes, part of the organic matter is absorbed and converted into microbial biomass, generically called biological or secondary sludge. This is mainly composed of biological solids, and for this reason it is also called a **biosolid**. The utilisation of this term still requires that the chemical and biological characteristics of the sludge are compatible with productive use, for example, in agriculture. The term 'biosolids' is a way of emphasising its beneficial aspects, giving more value to productive uses, in comparison with the mere non-productive final disposal by means of landfills or incineration.

The adequate final destination of biosolids is a fundamental factor for the success of a sanitation system. Nevertheless, this activity has been neglected in many developing countries. It is usual that in the design of wastewater treatment plants, the topic concerning sludge management is disregarded, causing this complex activity to be undertaken without previous planning by plant operators, and frequently under emergency conditions. Because of this, inadequate alternatives of final disposal have been adopted, largely reducing the benefits accomplished by the sewerage systems.

Although the sludge represents only 1% to 2% of the treated wastewater volume, its management is highly complex and has a cost usually ranging from 20% to 60% of the total operating costs of the wastewater treatment plant. Besides its economic importance, the final sludge destination is a complex operation, because it is frequently undertaken outside the boundaries of the treatment plant.

This part of the book intends to present an integrated view of all sludge management stages, including generation, treatment and final disposal. The sections also aim at reflecting the main sludge treatment and final disposal technologies potentially used in warm-climate regions, associated with the wastewater treatment processes described throughout the book.

The understanding of the various chapters in this part of the book depends on the knowledge of the introductory aspects and general overview, namely:

- introduction to sludge treatment and disposal
- relationships in sludge: solids levels, concentration and flow
- summary of the quantity of sludge generated in the wastewater treatment processes
- sludge treatment stages
- introduction to sludge thickening, stabilisation, dewatering, disinfection and final disposal

These topics are analysed again in this part of the book, at a more detailed level. The main topics covered are listed below.

Main topic	Items covered
Sewage sludge: characteristics and production	• Sludge production in wastewater treatment plants • Fundamental relationships among variables • Sludge production estimates • Mass balance in sludge treatment
Main sludge contaminants	• Metals • Pathogenic organisms • Organic contaminants • Discharge of effluents into public sewerage systems
Sludge stabilisation processes	• Anaerobic digestion • Aerobic digestion
Removal of the water content from sewage sludges	• Sludge thickening • Sludge conditioning • Drying bed • Centrifuge • Filter press • Belt press • Thermal drying
Pathogen removal	• Sludge disinfection mechanisms • Composting • Autothermal aerobic digestion • Alkaline stabilisation • Pasteurisation • Thermal drying
Assessment of alternatives for sludge management at wastewater treatment plants	• Trends on sludge management in some countries • Conditions to be analysed before assessing alternatives • Methodological approach for the selection of alternatives • Organisation of an assessment matrix • Sludge management at the wastewater treatment plant
Land disposal of sludge	• Beneficial uses of biosolids • Requirements and associated risks • Use and handling • Storage, transportation, application and incorporation • Land disposal without beneficial purposes: *landfarming* • Criteria and regulations in some countries
Main types of sludge transformation and disposal	• Thermal drying • Wet air oxidation • Incineration • Disposal in landfills
Environmental impact assessment and compliance monitoring of final sludge disposal	• Description of the activity from the environmental point of view • Alternatives of final sludge disposal • Potentially negative environmental impacts • Indicators and parameters for final sludge disposal monitoring • Programme for monitoring the impacts

2

Sludge characteristics and production

M. von Sperling, R.F. Gonçalves

2.1 SLUDGE PRODUCTION IN WASTEWATER TREATMENT SYSTEMS

The understanding of the concepts presented in this chapter depends on the previous understanding of the more introductory concepts of sludge management.

The amount of sludge produced in wastewater treatment plants, and that should be directed to the sludge processing units, can be expressed in terms of *mass* (g of total solids per day, dry basis) and *volume* (m^3 of sludge per day, wet basis). Section 2.2 details the methodology for mass and volume calculations. A simplified approach is assumed here, expressing sludge production on *per capita* and COD bases.

In biological wastewater treatment, part of the COD removed is converted into biomass, which will make up the biological sludge. Various chapters of this book show how to estimate the excess sludge production as a function of the COD or BOD removed from the wastewater. Table 2.1 presents, for the sake of simplicity, the mass of suspended solids wasted per unit of *applied* COD (or influent COD), considering typical efficiencies of COD removal from several wastewater treatment processes. For instance, in the activated sludge process – extended aeration – each

ISBN: 1 84339 166 X. Published by IWA Publishing, London, UK.

Table 2.1. Characteristics and quantities of sludge produced in various wastewater treatment systems

	Characteristics of the sludge produced and wasted from the liquid phase (directed to the sludge treatment stage)			
Wastewater treatment system	kgSS/ kgCOD applied	Dry solids content (%)	Mass of sludge (gSS/ inhabitant·d) (a)	Volume of sludge (L/ inhabitant·d) (b)
Primary treatment (conventional)	**0.35–0.45**	**2–6**	**35–45**	**0.6–2.2**
Primary treatment (septic tanks)	**0.20–0.30**	**3–6**	**20–30**	**0.3–1.0**
Facultative pond	**0.12–0.32**	**5–15**	**12–32**	**0.1–0.25**
Anaerobic pond – facultative pond				
• Anaerobic pond	0.20–0.45	15–20	20–45	0.1–0.3
• Facultative pond	0.06–0.10	7–12	6–10	0.05–0.15
• Total	**0.26–0.55**	–	**26–55**	**0.15–0.45**
Facultative aerated lagoon	**0.08–0.13**	**6–10**	**8–13**	**0.08–0.22**
Complete-mix aerated – sedim. pond	**0.11–0.13**	**5–8**	**11–13**	**0.15–0.25**
Septic tank + anaerobic filter				
• Septic tank	0.20–0.30	3–6	20–30	0.3–1.0
• Anaerobic filter	0.07–0.09	0.5–4.0	7–9	0.2–1.8
• Total	**0.27–0.39**	**1.4–5.4**	**27–39**	**0.5–2.8**
Conventional activated sludge				
• Primary sludge	0.35–0.45	2–6	35–45	0.6–2.2
• Secondary sludge	0.25–0.35	0.6–1	25–35	2.5–6.0
• Total	**0.60–0.80**	**1–2**	**60–80**	**3.1–8.2**
Activated sludge – extended aeration	**0.50–0.55**	**0.8–1.2**	**40–45**	**3.3–5.6**
High-rate trickling filter				
• Primary sludge	0.35–0.45	2–6	35–45	0.6–2.2
• Secondary sludge	0.20–0.30	1–2.5	20–30	0.8–3.0
• Total	**0.55–0.75**	**1.5–4.0**	**55–75**	**1.4–5.2**
Submerged aerated biofilter				
• Primary sludge	0.35–0.45	2–6	35–45	0.6–2.2
• Secondary sludge	0.25–0.35	0.6–1	25–35	2.5–6.0
• Total	**0.60–0.80**	**1–2**	**60–80**	**3.1–8.2**
UASB reactor	**0.12–0.18**	**3–6**	**12–18**	**0.2–0.6**
UASB + aerobic post-treatment (c)				
• Anaerobic sludge (UASB)	0.12–0.18	3–4	12–18	0.3–0.6
• Aerobic sludge (post-treatment) (d)	0.08–0.14	3–4	8–14	0.2–0.5
• Total	**0.20–0.32**	**3–4**	**20–32**	**0.5–1.1**

Notes:

- In the units with long sludge detention times (e.g., ponds, septic tanks, UASB reactors, anaerobic filters), all values include digestion and thickening (which reduce sludge mass and volume) occurring within the unit itself.

(a) Assuming 0.1 kgCOD/inhabitant·d and 0.06 kgSS/inhabitant·d

(b) Litres of sludge/inhabitant·d = [(gSS/inhabitant·d)/(dry solids (%))] × (100/1,000) (assuming a sludge density of 1,000 kg/m^3)

(c) Aerobic post-treatment: activated sludge, submerged aerated biofilter, trickling filter

(d) Aerobic sludge withdrawn from UASB tanks, after reduction of mass and volume through digestion and thickening that occur within the UASB reactor (the aerobic excess sludge entering the UASB is also smaller, because, in this case, the solids loss in the secondary clarifier effluent becomes more influential).

Sources: Qasim (1985), EPA (1979, 1987), Metcalf and Eddy (1991), Jordão and Pessoa (1995), Gonçalves (1996), Aisse *et al.* (1999), Chernicharo (1997), Gonçalves (1999)

kilogram of COD influent to the biological stage generates 0.50 to 0.55 kg of suspended solids (0.50 to 0.55 kgSS/kgCOD applied).

Considering that every inhabitant contributes approximately 100 gCOD/day (0.1 kgCOD/inhab·d), the per capita SS (suspended solids) contribution can be also estimated. In wastewater treatment processes in which physical mechanisms of organic matter removal prevail, there is no direct link between the solids production and the COD removal. In such conditions, Table 2.1 presents *per capita* SS productions based on typical efficiencies of SS removal in the various stages of the wastewater treatment solids.

The solids presented in Table 2.1 constitute the solids fraction of the sludge; the remainder is made up of plain water. The dry solids (total solids) concentration expressed in percentage is related to the concentration in mg/L (see Section 2.3). A 2%-dry-solids sludge contains 98% water; in other words, in every 100 kg of sludge, 2 kg correspond to dry solids and 98 kg are plain water.

The per capita daily volume of sludge produced is calculated considering the daily per capita load and the dry solids concentration of the sludge (see formula in Table 2.1 and Section 2.3).

In this part of the book, the expressions *dry solids*, *total solids* and *suspended solids* are used interchangeably, since most of the total solids in the sludge are suspended solids.

From Table 2.1, it is seen that among the processes listed, stabilisation ponds generate the smaller volume of sludge, whereas conventional activated sludge systems produce the largest sludge volume to be treated. The reason is that the sludge produced in the ponds is stored for many years in the bottom, undergoing digestion (conversion to water and gases) and thickening, which greatly reduce its volume. On the other hand, in the conventional activated sludge process, sludge is not digested in the aeration tank, because its residence time (sludge age) is too low to accomplish this.

Table 2.1 is suitable exclusively for preliminary estimates. It is important to notice that the mass and volumes listed in the table are related to the sludge that is directed to the treatment or processing stage. Section 2.2 presents the sludge quantities processed in each sludge treatment stage and in the final disposal.

2.2 SLUDGE CHARACTERISTICS AT EACH TREATMENT STAGE

Sludge characteristics vary as the sludge goes through several treatment stages. The major changes are:

- *thickening, dewatering*: increase in the concentration of total solids (dry solids); reduction in sludge volume
- *digestion*: decrease in the load of total solids (reduction of volatile suspended solids)

These changes can be seen in Table 2.2, which presents the solids load and concentration through the sludge treatment stages. Aiming at a better understanding,

Table 2.2. Sludge characteristics in each stage of the treatment process

	Sludge removed from the liquid phase		Thickened sludge			Digested sludge			Dewatered sludge			
Wastewater treatment system	Sludge mass (gSS/inhabitant·d)	Dry solids conc. (%)	Sludge mass (gSS/inhabitant·d)	Thickening process	Dry solids conc. (%)	Sludge mass (gSS/inhabitant·d)	Digestion process	Dry solids conc. (%)	Sludge mass (gSS/inhabitant·d)	Dewatering process	Dry solids conc. (%)	Per-capita volume (L/ inhabitant·d)
Primary treatment (conventional)	35–45	2–6	35–45	Gravity	4–8	25–28	Anaerobic	4–8	25–28	Drying bed	35–45	0.05–0.08
									25–28	Filter press	30–40	0.06–0.09
									25–28	Centrifuge	25–35	0.07–0.11
									25–28	Belt press	25–40	0.06–0.11
Primary treatment (septic tanks)	20–30	3–6	–	–	–	–	–	–	20–30	Drying bed	30–40	0.05–0.10
Facultative pond	20–25	10–20	–	–	–	–	–	–	20–25	Drying bed	30–40	0.05–0.08
Anaerobic pond – facultative pond												
• Anaerobic pond	20–45	15–20	–	–	–	–	–	–	20–45	Drying bed	30–40	0.05–0.14
• Facultative pond	6–10	7–12	–	–	–	–	–	–	6–10	Drying bed	30–40	0.015–0.03
• Total	26–55	–	–	–	–	–	–	–	26–55	Drying bed	30–40	0.06–0.17
Facultative aerated lagoon	8 –13	6–10	–	–	–	–	–	–	8–13	Drying bed	30–40	0.02–0.04
Complete-mix aerat. lagoon – sedim. pond	11–13	5–8	–	–	–	–	–	–	11–13	Drying bed	30–40	0.025–0.04
Septic tank + anaerobic filter												
• Septic tank	20–30	3–6	–	–	–	–	–	–	20–30	Drying bed	30–40	0.05–0.10
• Anaerobic filter	7–9	0.5–4,0	–	–	–	–	–	–	7–9	Drying bed	30–40	0.02–0.03
• Total	27–39	1.4–5.4	–	–	–	–	–	–	27–39	Drying bed	30–40	0.07–0.13
Conventional activated sludge												
• Primary sludge	35–45	2–6	35–45	Gravity	4–8	25–28	Anaerobic	4–8	–	–	–	–
• Secondary sludge	25–35	0.6–1	25–35	Gravity	2–3	16–22	Aerobic	1,5–4	–	–	–	–
				Flotation	2–5							
				Centrifuge	3–7							
• Mixed sludge	60–80	1–2	60–80	Gravity	3–7	38–50	Anaerobic	3–6	38–50	Drying bed	30–40	0.10–0.17
				Centrifuge	4–8					Filter press	25–35	0.11–0.20
										Centrifuge	20–30	0.13–0.25
										Belt press	20–25	0.15–0.25

(Continued)

Table 2.2 (*Continued*)

Wastewater treatment system	Sludge removed from the liquid phase		Thickened sludge			Digested sludge			Dewatered sludge			
	Sludge mass (gSS/inhabitant·d)	Dry solids conc. (%)	Sludge mass (gSS/inhabitant·d)	Thickening process	Dry solids conc. (%)	Sludge mass (gSS/inhabitant·d)	Digestion process	Dry solids conc. (%)	Sludge mass (gSS/inhabitant·d)	Dewatering process	Dry solids conc. (%)	Per-capita volume (L/ inhabitant·d)
Activated sludge – extended aeration	40–45	0.8–1.2	40–45	Gravity Flotation Centrifuge	2–3 3–6 3–6	–	–	–	40–45	Drying bed Filter press Centrifuge Belt press	25–35 20–30 15–20 15–20	0.11–0.17 0.13–0.21 0.19–0.29 0.19–0.29
High rate trickling filter												
• Primary sludge	35–45	2–6	35–45	Gravity	4–8	–	–	–	–	–	–	–
• Secondary sludge	20–30	1–2.5	20–30	Gravity	1–3	–	–	–	–	–	–	–
• Mixed sludge	55–75	1.5–4	55–75	Gravity	3–7	38–47	Anaerobic	3–6	38–47	Drying bed Filter press Centrifuge Belt press	30–40 25–35 20–30 20–25	0.09–0.15 0.10–0.18 0.12–0.22 0.14–0.22
Submerged aerated biofilter												
• Primary sludge	35–45	2–6	35–45	Gravity	4–8	25–28	Anaerobic	4–8	–	–	–	–
• Secondary sludge	25–35	0.6–1	25–35	Gravity Flotation Centrifuge	2–3 2–5 3–7	16–22	Aerobic	1.5–4	–	–	–	–
• Mixed sludge	60–80	1–2	60–80	Gravity Centrifuge	3–7 4–8	38–50	Anaerobic	3–6	38–50	Drying bed Filter press Centrifuge Belt press	30–40 25–35 20–30 20–25	0.10–0.17 0.11–0.20 0.13–0.25 0.15–0.25
UASB Reactor	12–18	3–6	–	–	–	–	–	–	12–18	Drying bed Filter press Centrifuge Belt press	30–45 25–40 20–30 20–30	0.03–0.06 0.03–0.07 0.04–0.09 0.04–0.09

UASB + activated sludge												
• Anaerobic sludge (UASB)	12–18	3–4	–	–	–	–	–	–	–	–	–	–
• Aerobic sludge (activated sludge) (*)	8–14	3–4	–	–	–	–	–	–	–	–	–	–
• Mixed sludge (*)	20–32	3–4	–	–	–	–	–	–	20–32	Drying bed	30–45	0.04–0.11
										Filter press	25–40	0.05–0.13
										Centrifuge	20–30	0.07–0.16
										Belt press	20–30	0.07–0.16
UASB + aerobic biofilm reactor												
• Anaerobic sludge (UASB)	12–18	3–4	–	–	–	–	–	–	–	–	–	–
• Aerobic sludge (aerobic reactor) (*)	6–12	3–4	–	–	–	–	–	–	–	–	–	–
• Mixed sludge (*)	18–30	3–4	–	–	–	–	–	–	18–30	Drying bed	30–45	0.04–0.10
										Filter press	25–40	0.045–0.12
										Centrifuge	20–30	0.06–0.15
										Belt press	20–30	0.06–0.15

Remarks:

- Expression of values on a daily basis does not imply that the sludge is removed, treated and disposed of every day.
- Solids capture in each stage of the sludge treatment has not been considered in the table. Non-captured solids are assumed to be returned to the system as supernatants, drained liquids and filtrates. Solids capture must be considered during mass balance computations and when designing each stage of the sludge treatment (solids capture percentage is the percentage of the influent solids load to a particular unit that leaves with the sludge, going to the next stage of solids treatment) (see Section 2.3·d).
- Solids are converted to gases and water during digestion process, which reduces the solids load. In the anaerobic digestion of the activated sludge and trickling filter sludge, the so-called secondary digester has the sole purpose of storage and solids – liquid separation, and do not remove volatile solids.
- Litres of sludge/inhabitant·d = [(gSS/inhabitant·d)/(dry solids (%))] × (100/1,050) (assuming 1050 kg/m^3 as the density of the dewatered sludge).

(*) Surplus aerobic sludge flows back to UASB, undergoing thickening and digestion with the anaerobic sludge.

Sources: Qasim (1985), Metcalf and Eddy (1991), Jordão and Pessôa (1995), Chernicharo (1997), Aisse *et al.* (1999), Gonçalves (1999)

the sludge load is shown on a *per-capita* basis. In the last column, the per-capita daily volume of sludge to be disposed of is presented.

Example 2.1

For a 100,000-inhabitant wastewater treatment plant composed by an UASB reactor, estimate the amount of sludge in each stage of its processing.

Solution:

(a) Sludge removed from the UASB reactor, to be directed to the sludge treatment stage

Tables 2.1 and 2.2 show that the per capita sludge mass production varies from 12 to 18 gSS/inhabitant·d, whereas the per capita volumetric production is around 0.2 to 0.6 L/inhabitant·d for sludge withdrawn from UASB reactors. Assuming intermediate values in each range, one has the following total sludge production to be processed:

SS load in sludge: 100,000 inhabitants × 15 g/inhabitant·d
= 1,500,000 gSS/d = 1,500 kgSS/d
Sludge flow: 100,000 inhabitants × 0.4 L/inhabitant·d = 40,000 L/d = 40 m^3/d

Should one wish to compute the sludge production as a function of the applied COD load, the following information from Table 2.1 could be used: (a) sludge mass production: 0.12 to 0.18 kgSS/kg applied COD; (b) per capita COD production: around 0.1 kgCOD/inhabitant·d. Assuming an intermediate value for the sludge production range:

Sludge SS load: 100,000 inhabitants × 0.1 kgCOD/inhabitant·d
× 0.15 kgSS/kgCOD = 1,500 kgSS/d

This value is identical to the one calculated above, based on the per-capita SS production.

(b) Dewatered sludge, to be sent to final disposal

The surplus sludge removed from UASB reactors is already thickened and digested, requiring only dewatering prior to final disposal as dry sludge.

In this example, it is assumed that the dewatering is accomplished in sludge drying beds. Table 2.2 shows that the per capita mass production of dewatered sludge remains in the range of 12 to 18 gSS/inhabitant·d, whereas the per capita volumetric production is reduced to the range of 0.03 to 0.06 L/inhabitant·d. Using average values, the total sludge production to be disposed of is:

SS load in sludge: 100,000 inhabitants × 15 g/inhabitant·d
= 1,500,000 gSS/d = 1,500 kgSS/d
Sludge flow: 100,000 inhabitants × 0.04 L/inhabitant·d = 4,000 L/d = 4 m^3/d

This is the volume to be sent for final disposal. Assuming a specific weight of 1.05, the total sludge mass (dry solids + water) to go for final disposal is 4 × 1.05 = 4.2 ton/d.

Example 2.2

For a 100,000-inhabitant conventional activated sludge plant compute the amount of sludge in each stage of the sludge treatment.

Solution:

(a) Sludge removed from the activated sludge system, to be directed to the sludge treatment stage

The activated sludge system produces primary and secondary sludge. The estimate of their production can be obtained from Tables 2.1 and 2.2:

Sludge mass production:

- Primary sludge: 35 to 45 gSS/inhabitant·d
- Secondary sludge: 25 to 35 gSS/inhabitant·d
- Mixed sludge (total production): 60 to 80 gSS/inhabitant·d

Sludge volume production:

- Primary sludge: 0.6 to 2.2 L/inhabitant·d
- Secondary sludge: 2.5 to 6.0 L/inhabitant·d
- Mixed sludge (total production): 3.1 to 8.2 L/inhabitant·d

Assuming average figures in each range:

Sludge mass production:

- Primary sludge: 100,000 inhabitants × 40 gSS/inhabitant·d = 4,000,000 gSS/d = 4,000 kgSS/d
- Secondary sludge: 100,000 inhabitants × 30 gSS/inhabitant·d = 3,000,000 gSS/d = 3,000 kgSS/d
- Mixed sludge (production total): 4,000 + 3,000 = 7,000 kgSS/.d

Sludge volume production:

- Primary sludge: 100,000 inhabitants × 1.5 L/inhabitant·d = 150,000 L/d = 150 m^3/d
- Secondary sludge: 100,000 inhabitants × 4.5 L/inhabitant·d = 450,000 L/d = 450 m^3/d
- Mixed sludge (production total): 150 + 450 = 600 m^3/d

(b) Thickened mixed sludge

The mass production of the mixed sludge remains unchanged after thickening (see Table 2.2), so:

Thickened sludge: 7,000 kgSS/d

(c) Digested mixed sludge

Volatile solids are partially removed by digestion, therefore reducing the total mass of dry solids. From Table 2.2, the production of anaerobically digested

Example 2.2 (Continued)

mixed sludge is between 38 and 50 gSS/inhabitant·d. Assuming an average figure:

$$\text{Mixed digested sludge: } 100{,}000 \text{ inhabitants} \times 45 \text{ gSS/inhabitant}\cdot\text{d} = 4{,}500{,}000 \text{ gSS/d} = 4{,}500 \text{ kgSS/d}$$

It should be noted that the total mass of solids is reduced from 7,000 kgSS/d to 4,500 kgSS/d.

(d) Dewatered mixed sludge

Sludge dewatering does not change the total solids load (see Table 2.2). Therefore, the total mass production is:

$$\text{Dewatered sludge} = 4{,}500 \text{ kgSS/d}$$

The sludge volume underwent large reductions in the dewatering and thickening processes. For a centrifuged dewatered sludge, Table 2.2 gives the per capita production of 0.13 to 0.25 L/inhabitant·d. Adopting an intermediate value of 0.20 L/inhabitant·d, one has:

$$\text{Dewatered sludge} = 100{,}000 \text{ inhabitants} \times 0.20 \text{ L/inhabitant}\cdot\text{d} = 20{,}000 \text{ L/d} = 20 \text{ m}^3\text{/d}$$

This is the sludge volume to be disposed of. It is seen that the final sludge production from the conventional activated sludge system is much larger than that from the UASB reactor (Example 2.1).

Note: For the sake of simplicity, in both examples the solids capture efficiency at each of the different sludge treatment stages was not taken into account. The solids capture efficiency adopted was 100%. For the concept of solids capture see Section 2.3.d.

2.3 FUNDAMENTAL RELATIONSHIPS IN SLUDGE

To express the characteristics of the sludge, as well as the production in terms of mass and volume, it is essential to have an understanding of some fundamental relationships. The following important items have been already presented:

- relationship between solid levels and water content
- expression of the concentration of dry solids
- relation between flow, concentration and load

Additional items covered in the current section are:

- total, volatile and fixed solids
- sludge density

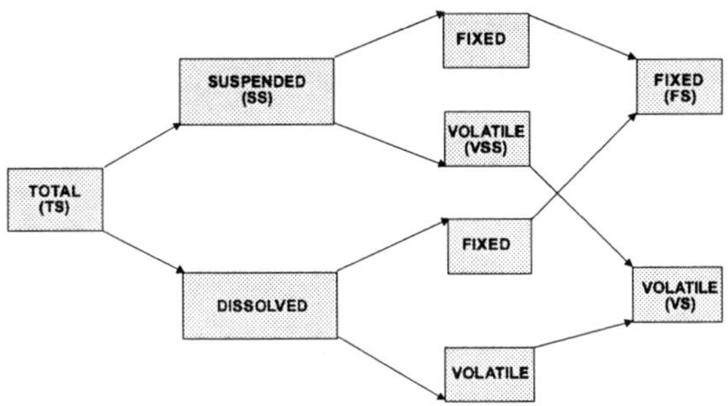

Figure 2.1. Sludge solids distribution according to size and organic fraction

- destruction of volatile solids
- solids capture

(a) Total, volatile and fixed solids

Sludge consists of solids and water. Total solids (TS) may be divided into suspended solids (SS) and dissolved solids. Most sludge solids are represented by suspended solids. Both suspended and dissolved solids may be split into inorganic or fixed solids (FS) and organic or volatile solids (VS). Figure 2.1 illustrates the distribution of the solids according to these different forms.

The ratio of volatile to total solids (VS/TS) gives a good indication of the organic fraction in the sludge solids, as well as its level of digestion. VS/TS ratio for undigested sludges ranges from 0.75 to 0.80, whereas for digested sludges the range is from 0.60 to 0.65. Table 2.3 presents typical ranges of VS/TS for sludges from different wastewater treatment processes.

In this part of the book, when calculating the solids load along the sludge treatment line, the expressions *dry solids*, *total solids* and even *suspended solids* (admitting that the majority of total solids of the sludge is suspended solids) are being used interchangeably.

(b) Density and specific gravity of the sludge

The specific gravity of the fixed solids particles is approximately 2.5 (Crites and Tchobanoglous, 2000), whereas for volatile solids the specific gravity is approximately 1.0. For water, the value is, of course, 1.0. The density of the sludge (water plus solids) depends upon the relative distribution among those three components.

The specific gravity of the *sludge solids* can be estimated by (Metcalf and Eddy, 1991; Crites and Tchobanoglous, 2000):

$$\text{Specific gravity of solids} = \frac{1}{\left(\frac{(\text{FS/TS})}{2.5} + \frac{(\text{VS/TS})}{1.0}\right)} \quad (2.1)$$

Table 2.3. Density, specific gravity, VS/TS ratio and percentage of dry solids for various sludge types

Types of sludge	VS/ST Ratio	% dry solids	Specific gravity of solids	Specific gravity of sludge	Density of sludge (kg/m^3)
Primary sludge	0.75–0.80	2–6	1.14–1.18	1.003–1.01	1003–1010
Secondary anaerobic sludge	0.55–0.60	3–6	1.32–1.37	1.01–1.02	1010–1020
Secondary aerobic sludge (conv. AS)	0.75–0.80	0.6–1.0	1.14–1.18	1.001	1001
Secondary aerobic sludge (ext. aer.)	0.65–0.70	0.8–1.2	1.22–1.27	1.002	1002
Stabilisation pond sludge	0.35–0.55	5–20	1.37–1.64	1.02–1.07	1020–1070
Primary thickened sludge	0.75–0.80	4–8	1.14–1.18	1.006–1.01	1006–1010
Second thickened sludge (conv. AS)	0.75–0.80	2–7	1.14–1.18	1.003–1.01	1003–1010
Second thickened sludge (ext. aer.)	0.65–0.70	2–6	1.22–1.27	1.004–1.01	1004–1010
Thickened mixed sludge	0.75–0.80	3–8	1.14–1.18	1.004–1.01	1004–1010
Digested mixed sludge	0.60–0.65	3–6	1.27–1.32	1.007–1.02	1007–1020
Dewatered sludge	0.60–0.65	20–40	1.27–1.32	1.05–1.1	1050–1100

Notes:
For specific gravity of solids use *Equation* 2.1; for specific gravity of sludge use *Equation* 2.2
AS = activated sludge; ext. aer. = extended aeration activated sludge

On its turn, the specific gravity of the *sludge* (water plus solids) can be estimated as follows:

specific gravity of sludge

$$= \frac{1}{\left(\dfrac{\text{Solids fraction in sludge}}{\text{Sludge density}} + \dfrac{\text{Water fraction in sludge}}{1.0}\right)} \qquad (2.2)$$

The solids fraction in the sludge corresponds to the dry solids (total solids), expressed in decimals, whereas the water fraction in the sludge corresponds to the moisture, also expressed in decimals (and not in percentage).

Applying the above relationships, one obtains the density and specific gravity of solids and sludges presented in Table 2.3, for different types of sludges.

Table 2.3 shows that the sludge densities are very close to the water density. Nevertheless, it should be noted that some authors indicate slightly higher densities than those from Table 2.3, which have been computed following the above procedure. Usual values reported are presented in Table 2.4.

(c) Destruction of volatile solids

Digestion removes biodegradable organic solids from the sludge. Hence, it can be said that there was a removal or destruction of volatile solids (VS). The quantity

Table 2.4. Usual values of sludge densities

Type of sludge	Specific gravity	Density (kg/m^3)
Primary sludge	1.02–1.03	1020–1030
Secondary anaerobic sludge	1.02–1.03	1020–1030
Secondary aerobic sludge	1.005–1.025	1005–1025
Thickened sludge	1.02–1.03	1020–1030
Digested sludge	1.03	1030
Dewatered sludge	1.05–1.08	1050–1080

of fixed solids (FS) remains unchanged. Typical efficiencies of VS removal in digestion are:

$$\boxed{E = 0.40 \text{ to } 0.55 \text{ (40 to 55\%)}}$$

The solids load (kg/d) before and after digestion can be computed from:

$$\boxed{TS_{influent} = VS_{influent} + FS_{influent}} \quad (2.3)$$

$$\boxed{TS_{effluent} = (1 - E) \times VS_{influent} + FS_{influent}} \quad (2.4)$$

(d) Solids capture

In the sludge treatment stages in which there is solids–liquid separation (e.g., thickening and dewatering), not all solids are separated from the liquid and go to the subsequent stage of the sludge treatment. A part of these solids remain in the supernatants, drained outflows and filtrates of the separation units. Because of these remaining solids (particulate BOD), these flows must be returned to the head of the works to be mixed with the plant influent and undergo additional treatment.

The incorporation of solids to sludge is known as *solids capture* (or *solids recovery*). It is expressed usually as a percentage (%), aiming to depict the efficiency of incorporation of solids to the sludge that will be sent to the subsequent stages of the processing.

Therefore, the solids loads (kgSS/d) are:

$$\boxed{\text{Effluent SS load in sludge} = \text{Solids capture} \times \text{Influent SS load in sludge}} \quad (2.5)$$

$$\boxed{\text{SS load in drained liquid} = (1 - \text{Solids capture}) \times \text{Influent SS load in sludge}} \quad (2.6)$$

For example, if a SS load of 100 kgSS/d goes through a 90% solids capture efficiency sludge treatment unit, then 90 kgSS/d (= 0.9 × 100 kgSS/d) will flow with the sludge towards the subsequent treatment stages, and 10 kgSS/d (= (1 − 0.9) × 100 kgSS/d) will be incorporated to the drained liquid and be sent back to the head of the wastewater treatment plant.

Typical values of solids capture in sludge treatment are presented in Table 2.5.

Table 2.5. Ranges of solids captures in sludge treatment

Type of sludge	Thickening		Digestion		Dewatering	
	Process	Capture (%)	Process	Capture (%)	Process	Capture (%)
Primary sludge	Gravity	85–92	Second. digester	95	Drying bed	90–98
					Filter press	90–98
					Centrifuge	90–95
					Belt press	90–95
Secondary sludge	Gravity	75–85	Second. digester	90–95	Drying bed	90–98
	Flotation	80–95			Filter press	90–98
	Centrifuge	80–95			Centrifuge	90–95
					Belt press	90–95
Mixed sludge	Gravity	80–90	Second. digester	90–95	Drying bed	90–98
	Centrifuge	85–95			Filter press	90–98
					Centrifuge	90–95
					Belt press	90–95

Note: The *secondary* anaerobic digester merely works as a sludge holder and solids–liquid separator. The *primary* anaerobic digester has 100% solids capture, because all solids (as well as liquid) are sent to the secondary digester. The *aerobic* digester has also 100% capture, with no further storage stage.
Source: Adapted from Qasim (1985) and EPA (1987)

2.4 CALCULATION OF THE SLUDGE PRODUCTION

2.4.1 Primary sludge production

The sludge production in primary treatment (primary sludge) depends on the SS removal efficiency in the primary clarifiers. This efficiency can be also understood as solids capture. Typical SS removal (capture) efficiencies in *primary clarifiers* are as follows:

SS removal efficiency in primary clarifiers: E = 0.60 to 0.65 (60 to 65%)

Therefore, the load of primary sludge produced is:

$$\text{SS load from primary sludge} = \text{E} \times \text{Influent SS load}$$

$$\text{SS load from primary sludge} = \text{E. Q. Influent SS conc} \quad (2.7)$$

The SS load direct to the biological treatment is:

$$\text{Influent SS load to biological treatment} = (1 - \text{E}).\text{Q}.\text{Influent SS conc} \quad (2.8)$$

The volumetric production of the primary sludge can be estimated from Equation 5.5, and the TS concentration and specific gravity of the sludge from Table 2.4.

Example 2.4 shows an estimate of primary sludge production, as well as the transformations in sludge load and volume that take place throughout the various sludge treatment units.

2.4.2 Secondary sludge production

Secondary (biological) sludge production is estimated considering kinetic and stoichiometric coefficients of the particular biological wastewater treatment process being used. The following fractions make up the sludge produced:

- Biological solids: biological solids produced in the system as a result of the organic matter removal.
- Inert solids from raw sewage: non-biodegradable solids, accumulated in the system.

The *net* production of biological solids corresponds to the *total* production (synthesis, or anabolism) minus *mortality* (decay, or catabolism).

Various chapters in this book present an estimate of the total sludge production in their respective wastewater treatment process following the preceding methodology. Therefore, further details should be obtained in these chapters. Approximate figures for sludge productions can be derived from Tables 2.1 and 2.2.

In the estimation of the amount of biological sludge to be treated, a fraction may be deducted from the total amount produced. This fraction corresponds to the amount lost with the final effluent (solids that unintentionally escape with the final effluent, due to the fact that the SS removal efficiencies are naturally lower than 100% in the final clarifiers). If this refinement in the calculation is incorporated, it should be understood that the load of solids to be treated is equal to the load of solids produced minus the load of solids escaping with the final effluent.

Example 2.3 shows the estimation of the sludge production from an UASB reactor, whereas Example 2.4 computes the primary and secondary sludge production from an activated sludge system. Sludge load and volume variations along the sludge treatment are also quantified in both examples.

Example 2.3

Estimate the sludge flow and concentration and the SS load in each stage of the sludge processing at a treatment plant composed by an UASB reactor, treating the wastewater from 20,000 inhabitants.

The sludge treatment flowsheet is made up of:

- Type of sludge: secondary sludge (withdrawn from the UASB reactor)
- Sludge dewatering: natural (drying beds)

Data:

- Population: 20,000 inhabitants
- Average influent flow: $Q = 3{,}000\ m^3/d$
- Concentration of influent COD: $S_o = 600$ mg/L
- Solids production coefficient: $Y = 0.18\ kgSS/kgCOD_{applied}$
- Expected concentration of the excess sludge: 4%
- Sludge density: 1020 kg/m^3

***Example 2.3* (*Continued*)**

Solution:

(a) Sludge generated in the UASB reactor (influent to the dewatering stage)

$$\begin{aligned}\text{COD load applied} &= 3{,}000\ \text{m}^3\text{/d} \times 600\ \text{g/m}^3\\ &= 1{,}800{,}000\ \text{gCOD/d} = 1{,}800\ \text{kgCOD/d}\\ \text{Sludge production: P} &= 0.18\ \text{kgSS/kgCOD}_{\text{applied}} \times 1{,}800\ \text{kgCOD/d}\\ &= 324\ \text{kgSS/d}\end{aligned}$$

Sludge flow:

$$\text{Sludge flow (m}^3\text{/d)} = \frac{\text{SS load (kgSS/d)}}{\dfrac{\text{Dry solids (\%)}}{100} \times \text{Sludge density (kg/m}^3\text{)}}$$

$$= \frac{324\ \text{kgSS/d}}{\dfrac{4}{100} \times 1{,}020\ \text{kg/m}^3} = 7.94\ \text{m}^3\text{/d}$$

This is the same value obtained in the referred to example.

The per capita productions are:

- Per capita SS load = 324 kgSS/d/20,000 inhabitants = 16 gSS/inhabitant·d
- Per capita flow = 7.94 m^3/d/20,000 inhabitants = 0.40 L/inhabitant·d

These values are within the per capita ranges presented in Table 2.1.

(b) Effluent sludge from dewatering (sludge for final disposal)

Since the excess sludge from the UASB reactor is already digested and thickened, only dewatering before final disposal is required.

In case the sludge is dewatered using drying beds, its dry solids content is between 30% to 45% (see Table 2.2), its density is in the range from 1050 to 1080 kg/m^3 (Table 2.4) and the solids capture is between 90% to 98% (see Table 2.5). In this example, the following values are adopted:

- SS concentration in the dewatered sludge: 40%
- density of the dewatered sludge: 1,060 kg/m^3
- solids capture in the dewatering stage: 95%

The solids captured and incorporated to the dewatered sludge can be calculated from Equation 2.5:

$$\begin{aligned}\text{Effluent SS load (kgSS/d)} &= \text{Solids capture} \times \text{SS influent load (kgSS/d)}\\ &= 0.95 \times 324\ \text{kgSS/d} = 308\ \text{kgSS/d}\end{aligned}$$

Example 2.3 (Continued)

The daily volume of dewatered sludge (cake) to go for final disposal can be estimated by:

$$\text{Sludge flow (m}^3\text{/d)} = \frac{\text{SS load (kgSS/d)}}{\dfrac{\text{Dry solids (\%)}}{100} \times \text{Sludge density (kg/m}^3\text{)}}$$

$$= \frac{308 \text{ kgSS/d}}{\dfrac{40}{100} \times 1060 \text{ kg/m}^3} = 0.73 \text{ m}^3\text{/d}$$

The per capita productions are:

- Per capita SS load = 308 kgSS/d/20,000 inhabitants = 15.4 gSS/inhabitant·d
- Per capita flow = 0.73 m^3/d/20,000 inhabitants = 0.04 L/inhabitant·d

These values are within the per capita ranges presented in Table 2.2.

(c) Filtrate from dewatering (returned to the head of the WWTP)

The solids load that is incorporated to the drying bed filtrate liquid and returns to the head of the WWTP may be computed from Equation 2.6:

$$\text{SS load in filtrate (kgSS/d)} = (1 - \text{Solids capture}) \times \text{Influent SS load (kgSS/d)}$$
$$= (1 - 0.95) \times 324 \text{ kgSS/d} = 16 \text{ kgSS/d}$$

The flow of the filtrate from the drying beds (without consideration of evaporation, for the sake of simplicity in this example) is the difference between the influent and effluent sludge flows:

$$\text{Filtrate flow} = \text{Influent sludge flow} - \text{Effluent sludge flow}$$
$$= 7.94 - 0.73 = 7.21 \text{ m}^3\text{/d}$$

The filtrate solids concentration is the SS load divided by the filtrate flow (the filtrate and water densities are assumed to be equal):

$$\text{SS conc} = \frac{\text{SS load}}{\text{Flow}} = \frac{16 \text{ kgSS/d} \times 1{,}000 \text{ g/kg}}{7.21 \text{ m}^3\text{/d}}$$
$$= 2{,}219 \text{ g/m}^3 = 2{,}219 \text{ mg/L} = 0.22\%$$

The preceding solids load can be taken into account in the computation of the influent load to the UASB reactor.

Example 2.4

Estimate the sludge flow and concentration and the SS load in each stage of the sludge processing at a treatment plant composed by a conventional activated sludge plant, treating the wastewater from 62,000 inhabitants.

The sludge treatment flowsheet is made up of:

- Types of sludge: primary and secondary (mixed when entering the sludge treatment)
- Type of sludge thickening: gravity
- Type of sludge digestion: primary and secondary anaerobic digesters
- Type of sludge dewatering: mechanical (centrifuge)

Pertinent data from the referred to example:

- Population: 67,000 inhabitants
- Average influent flow: $Q = 9{,}820\ m^3/d$
- Influent SS load: 3,720 kg/d
- Influent SS concentration: SS = 379 mg/L
- SS removal efficiency in the primary clarifier: 60% (assumed)

Data related to the production of secondary sludge (from the referred to example):

- Place of removal of excess sludge: return sludge line
- SS load to be removed: 1,659 kgSS/d
- SS concentration in excess sludge: 7,792 mg/L (0.78%)
- Excess sludge flow: $Q_{ex} = 213\ m^3/d$

Solution:

(a) Sludge removed from the primary clarifier (primary sludge)

SS load removed from primary clarifier:

$$\begin{aligned}\text{Removed SS load} &= \text{Removal efficiency} \times \text{Influent SS load}\\ &= 0.60 \times 3{,}720\ \text{kgSS/d} = 2{,}232\ \text{kgSS/d}\end{aligned}$$

The characteristics of the removed primary sludge are: dry solids content from 2% to 6% (see Tables 2.2 and 2.3) and sludge density from 1020 to 1030 kg/m^3 (Table 2.3). The values adopted for the present example are:

- SS concentration in primary sludge: 4%
- Primary sludge density: 1020 kg/m^3

Example 2.4 (Continued)

The flow of primary sludge that goes for thickening is estimated by:

$$\text{Sludge flow (m}^3\text{/d)} = \frac{\text{SS load (kgSS/d)}}{\frac{\text{Dry solids (\%)}}{100} \times \text{Sludge density (kg/m}^3\text{)}}$$

$$= \frac{2{,}232 \text{ kgSS/d}}{\frac{40}{100} \times 1020 \text{ kg/m}^3} = 54.7 \text{ m}^3\text{/d}$$

The per capita primary sludge productions are:

- Per capita SS load = 2,232 kgSS/d/67,000 inhabitants = 33 gSS/inhabitant·d
- Per capita sludge flow = 54.7 m^3/d/67,000 inhabitants = 0.82 L/inhabitant·d

These values are within the lower range of per capita values presented in Table 2.1.

(b) Secondary sludge

The amount of secondary sludge to be removed from the activated sludge system was calculated and it is now an input data for the present example (see above):

- Place of removal of excess sludge: return sludge line
- SS load to be removed: 1,659 kgSS/d
- SS concentration in excess sludge: 7,792 mg/L (0.78%)
- Excess sludge flow: $Q_{ex} = 213$ m^3/d

The per capita secondary sludge productions are:

- Per capita SS load = 1,659 kgSS/d/67,000 inhabitants = 25 gSS/inhabitant·d
- Per capita sludge flow = 213 m^3/d/67,000 inhabitants = 3.18 l/inhabitant·d

These values are within the lower range of per capita values of Table 2.1

(c) Mixed sludge (primary sludge + secondary sludge) (influent sludge to the thickener)

Primary and secondary sludges are mixed before entering the thickener.

SS load in mixed sludge is:

$$\text{Mixed sludge SS load} = \text{Primary sludge SS load} + \text{Secondary sludge SS load}$$
$$= 2{,}232 + 1{,}659 = 3{,}891 \text{ kgSS/d}$$

***Example 2.4* (*Continued*)**

The mixed sludge flow is:

$$\text{Mixed sludge flow} = \text{Primary sludge flow} + \text{Secondary sludge flow}$$
$$= 54.7 + 213.0 = 267.7\ \text{m}^3/\text{d}$$

The solids concentration in the mixed sludge is the SS load divided by the sludge flow (considering the mixed sludge density equal to the water density):

$$\text{SS conc} = \frac{\text{SS load}}{\text{Flow}} = \frac{3{,}891\ \text{kgSS/d} \times 1{,}000\ \text{g/kg}}{267.7\ \text{m}^3/\text{d}}$$
$$= 14{,}535\ \text{g/m}^3 = 14{,}535\ \text{mg/L} = 1.45\%$$

(d) Thickened effluent sludge (sludge to be sent to the digester)

The effluent sludge from the thickener has a solids load equal to the influent load multiplied by the solids capture. From Table 2.5, it is seen that the solids capture for gravity thickening of primary plus secondary sludge is between 80% and 90%. Assuming 85% solids capture, the effluent SS load from the thickener is (Equation 2.5):

$$\text{SS effluent load} = \text{Solids capture} \times \text{Influent load}$$
$$= 0.85 \times 3{,}891\ \text{kg/d} = 3{,}307\text{kgSS/d}$$

The mixed sludge thickened by gravity has the following characteristics: dry solids content between 3% and 7% (see Table 2.2), and sludge density from 1,020 to 1,030 kg/m^3 (see Table 2.4). The following values are adopted in the present example:

- SS concentration in thickened sludge: 5%
- Density of thickened sludge: 1,030 kg/m^3

The thickened sludge flow going to digestion is estimated by:

$$\text{Sludge flow (m}^3\text{/d)} = \frac{\text{SS load (kgSS/d)}}{\dfrac{\text{Dry solids (\%)}}{100} \times \text{Sludge density (kg/m}^3\text{)}}$$
$$= \frac{3{,}307\ \text{kgSS/d}}{\dfrac{5}{100} \times 1{,}030\ \text{kg/m}^3} = 64.2\ \text{m}^3/\text{d}$$

(e) Thickener supernatant (returned to the head of the treatment plant)

The SS load in the thickener supernatant is:

$$\text{Supernatant SS load} = \text{Influent SS load} - \text{Effluent sludge SS load}$$
$$= 3{,}891 - 3{,}307 = 584\ \text{kgSS/d}$$

Example 2.4 (Continued)

The thickener supernatant flow is:

$$\text{Supernatant flow} = \text{Influent flow} - \text{Effluent sludge flow}$$
$$= 267.7 - 64.2 = 203.5 \text{m}^3/\text{d}$$

The SS concentration in the supernatant is:

$$\text{SS conc} = \frac{\text{SS load}}{\text{Flow}} = \frac{584 \text{ kgSS / d} \times 1{,}000 \text{ g/kg}}{203.5 \text{ m}^3/\text{d}}$$
$$= 2{,}870 \text{ g/m}^3 = 2{,}870 \text{ mg/L} = 0.29\%$$

(f) Effluent sludge from primary digester (influent sludge to secondary digester)

The VS/TS ratio in the thickened mixed sludge is between 0.75 to 0.80 (see Table 2.3). In the present example, the value of 0.77 has been adopted. The distribution of the TS influent load to digestion, considering 77% as volatile solids and 23% as fixed solids, is:

- TS influent = 3,307 kgTS/d
- VS influent = (VS/TS) × TS influent = 0.77 × 3,307 = 2,546 kgVS/d
- FS influent = (1 − VS/TS) × TS influent = (1 − 0.77) × 3,307 = 761 kgFS/d

After digestion, the FS remain unaltered, but the VS are partially removed. According to Section 2.3.c, the removal efficiency of VS in anaerobic digesters is between 40% and 55%. For the present example, 50% (0.50) removal was assumed.

The distribution of the solids load from the effluent from the primary digester is:

- FS effluent = FS influent = 761 kgFS/d
- VS effluent = (1 − VS removal efficiency) × VS influent = (1 − 0.50) × 2,546 = 1,273 kgVS/d
- TS effluent = FS effluent + VS effluent = 761 + 1273 = 2,034 kgTS/d

The distribution of effluent solids from the primary digester is:

- FS/TS effluent = 761/2,034 = 0.37 = 37%
- VS/TS effluent = 1,273/2,034 = 0.63 = 63%

It should be noticed that the VS/TS ratio (77%) in the influent to the digester has been reduced down to 63% after digestion.

The effluent sludge flow from the primary digester is equal to the influent sludge flow. Therefore:

$$\text{Primary digester effluent sludge flow} = \text{Primary digester influent sludge flow}$$
$$= 64.2 \text{ m}^3/\text{d}$$

Example 2.4 (Continued)

The SS concentration in the effluent sludge from the primary digester is:

$$SS\ conc = \frac{SS\ load}{Flow} = \frac{2{,}034\ kgSS/d\ \times 1{,}000\ g/kg}{64.2\ m^3/\ d}$$
$$= 31{,}682\ g/m^3 = 31{,}682\ mg/L = 3.17\%$$

It can be seen that the digestion process lead to a reduction of both the solids load and the solids concentration.

(g) Effluent sludge from the secondary digester (sludge to be dewatered)

The secondary digester does not actually digest solids, being simply a sludge holding tank. During the sludge storage, some sedimentation of solids takes place. A supernatant is formed and removed, being returned to the head of the works. The sludge settled in the bottom proceeds to dewatering.

The solids capture in the secondary digester is between 90% and 95% (see Table 2.5). Assuming a solids capture of 95%, the effluent SS load from the secondary digester is:

$$Effluent\ SS\ load = Solids\ capture \times Influent\ load$$
$$= 0.95 \times 2{,}034\ kg/d = 1{,}932\ kgSS/d$$

The volatile and fixed solids keep the same relative proportions they had when leaving the primary digester (FS/TS = 37%; VSS/TS = 63%, as computed in item f). The effluent VS and FS loads from the secondary digester are:

- FS effluent load = 0.37 × 1,932 = 715 kgFS/d
- VS effluent load = 0.63 × 1,932 = 1,217 kgVS/d

The mixed digested sludge has the following characteristics: dry solids content between 3% and 6% (see Table 2.2), and sludge density around 1030 kg/m^3 (see Table 2.4). The following values are adopted in the present example:

- SS concentration in the effluent sludge from the secondary digester: 4% (this figure must be higher than the SS concentration in the effluent sludge from the primary digester, which was 3.17% in this particular example)
- Density of the effluent sludge from the secondary digester: 1,030 kg/m^3

The effluent sludge flow from the secondary digester sent to dewatering is estimated by:

$$Sludge\ flow\ (m^3/d) = \frac{SS\ load\ (kgSS/d)}{\dfrac{Dry\ solids\ (\%)}{100} \times Sludge\ density\ (kg/m^3)}$$
$$= \frac{1{,}932\ kgSS/d}{\dfrac{4}{100} \times 1{,}030\ kg/m^3} = 46.9\ m^3/d$$

Example 2.4 (Continued)

(h) Supernatant from the secondary digester (returned to the head of the treatment plant)

The SS load in the secondary digester supernatant is:

$$\text{Supernatant SS load} = \text{Influent SS load} - \text{Effluent SS sludge load}$$
$$= 2{,}034 - 1{,}932 = 102 \text{ kgSS/d}$$

The digester supernatant flow is:

$$\text{Supernatant flow} = \text{Influent sludge flow} - \text{Effluent sludge flow}$$
$$= 64.2 - 46.9 = 17.3 \text{ m}^3\text{/d}$$

The SS concentration in the supernatant is:

$$\text{SS conc} = \frac{\text{SS load}}{\text{Flow}} = \frac{102 \text{ kgSS/d} \times 1{,}000 \text{ g/kg}}{17.3 \text{ m}^3\text{/d}}$$
$$= 5{,}896 \text{ g/m}^3 = 5{,}896 \text{ mg/L} = 0.59\%$$

(i) Dewatered sludge production (sludge for final disposal)

In the present example, dewatering is accomplished by centrifuges. The solids load due to the polyelectrolytes added to the sludge being centrifuged is not taken into account. It is assumed that the dewatered sludge sent for final disposal does not receive any other chemicals (for instance, lime for disinfection). If lime is added, its solids load is significant and should be taken into consideration (see Chapter 6).

The solids load in the dewatered sludge (sludge cake) is equal to the influent load multiplied by the solids capture. According to Table 2.5, the capture of digested mixed sludge solids through centrifuge dewatering is from 90% to 95%. Assuming 90% solids capture, the effluent SS load from the dewatering stage is (Equation 2.5):

$$\text{SS effluent load} = \text{Solids capture} \times \text{Influent load}$$
$$= 0.90 \times 1{,}932 \text{ kg/d} = 1{,}739 \text{ kgSS/d}$$

The mixed sludge dewatered by centrifuges has the following characteristics: dry solids content between 20% and 30% (see Table 2.2) and sludge density between 1,050 and 1,080 kg/m^3 (see Table 2.4). The following values are adopted in the present example:

- SS concentration in the dewatered sludge: 25%
- Density of the dewatered sludge: 1,060 kg/m^3

Example 2.4 (Continued)

The daily volume (flow) of dewatered sludge sent for final disposal is estimated by:

$$\text{Sludge flow (m}^3\text{/d)} = \frac{\text{SS load (kgSS/d)}}{\dfrac{\text{Dry solids (\%)}}{100} \times \text{Sludge density (kg/m}^3\text{)}}$$

$$= \frac{1{,}739 \text{ kgSS/d}}{\dfrac{25}{100} \times 1{,}060 \text{ kg/m}^3} = 6.6 \text{ m}^3\text{/d}$$

The per capita production of mixed dewatered sludge is:

- Per capita SS load = 1,739 kgSS/d/67,000 inhabitants = 26 gSS/inhabitant·d
- Per capita flow = 6.6 m^3/d/67,000 inhabitants = 0.10 L/inhabitant·d

These values are below the per capita figures of Table 2.2. However, Table 2.2 does not consider the solids capture efficiency, and assumes 100% capture in each one of the various steps of the sludge treatment, that is, all the influent sludge leaves in the effluent to the next stage of treatment. On the other hand, the present example did not consider the supernatant load, neither the drained solids load (both figures have been computed, but not added as further influent loads to the WWTP). Section 2.5 exemplifies how such returned loads can be incorporated to the general plant mass balance.

(j) Centrate from dewatering (returned to head of the treatment plant)

The SS load present in the centrifuge drained flow (centrate) is:

$$\text{Drained SS load} = \text{Influent SS load} - \text{Effluent sludge SS load}$$
$$= 1{,}932 - 1{,}739 = 193 \text{ kgSS/d}$$

The centrifuge drained flow is:

$$\text{Drained flow} = \text{Influent flow} - \text{Effluent sludge flow}$$
$$= 46.9 - 6.6 = 40.3 \text{ m}^3\text{/d}$$

The SS concentration in the drained liquid is:

$$\text{SS conc} = \frac{\text{SS load}}{\text{Flow}} = \frac{193 \text{ kgSS/d} \times 1{,}000 \text{ g/kg}}{40.3 \text{ m}^3\text{/d}}$$
$$= 4{,}789 \text{ g/m}^3 = 4{,}789 \text{ mg/L} = 0.48\%$$

Example 2.4 (Continued)

(k) Summary of loads, flows and concentrations

Source	Sludge SS load (kgSS/d)	Sludge Flow (m^3/d)	Sludge SS concent. (%)	Supernatant/drained SS load (kgSS/d)	Supernatant/drained Flow (m^3/d)	Supernatant/drained SS concent. (mg/L)
Primary sludge	2,232	54.7	4.00	–	–	–
Secondary sludge	1,659	213.0	0.78	–	–	–
Mixed sludge	3,891	267.7	1.45	–	–	–
Thickener	3,307	64.2	5.0	584	203.5	2870
Primary digester	2,034	64.2	3.2	–	–	–
Secondary digester	1,932	46.9	4.0	102	17.3	5896
Dewatering	1,739	6.6	25.0	193	40.3	4789

The following formulae allow the structuring of the table in a spreadsheet format:

Source	Sludge SS load (kgSS/d)	Sludge Flow (m^3/d)	Sludge SS concent. (%)	Supernatant/drained SS load (kgSS/d)	Supernatant/drained Flow (m^3/d)	Supernatant/drained SS concent. (mg/L)
Primary sludge	(1)	(2)	(3)	–	–	–
Secondary sludge	(4)	(5)	(6)	–	–	–
Mixed sludge	(7)	(8)	(9)	–	–	–
Thickener	(10)	(11)	(12)	(13)	(14)	(15)
Primary digester	(16)	(17)	(18)	–	–	–
Secondary digester	(19)	(20)	(21)	(22)	(23)	(24)
Dewatering	(25)	(26)	(27)	(28)	(29)	(30)

(1) = Clarifier solids capture efficiency × Influent SS load to the primary clarifier

$$(2) = \frac{(1)}{\frac{(3)}{100} \times \text{Density (kg/m}^3)}$$

(3) = Assumed value

(4) (5) (6) = Calculated values based on activated sludge process kinetics

(7) = (1) + (4)

(8) = (2) + (5)

(9) = [(7) × 100]/[(8) × 1000]

(10) = Thickener solids capture efficiency × (7)

$$(11) = \frac{(10)}{\frac{(12)}{100} \times \text{Density (kg/m}^3)}$$

(12) = Assumed value

Example 2.4 (Continued)

(13) = (7) − (10)

(14) = (8) − (11)

(15) = (13) × 1000/(14)

(16) = [(10) × (1 − VSS/SS sludge)] + [(10) × (VSS/SS sludge) × VSS removal efficiency)]

(17) = (11)

(18) = [(16) × 100]/[(17) × 1000]

(19) = Secondary digester solids capture efficiency × (16)

$$(20) = \frac{(19)}{\frac{(21)}{100} \times \text{Density (kg/m}^3\text{)}}$$

(21) = Assumed value

(22) = (16) − (19)

(23) = (17) − (20)

(24) = (22) × 1000/(23)

(25) = Dewatering solids capture efficiency × (19)

$$(26) = \frac{(25)}{\frac{(27)}{100} \times \text{Density (kg/m}^3\text{)}}$$

(27) = Assumed value

(28) = (19) − (25)

(29) = (20) − (26)

(30) = (28) × 1000/(29)

2.5 MASS BALANCE IN SLUDGE TREATMENT

As seen on Examples 2.3 and 2.4, supernatant, percolated and drained liquids from the various sludge treatment stages contain suspended solids, since not all influent solids are able to come out with the sludge, because the solids capture efficiency is not 100%. These solids represent organic matter and must return to the sludge treatment plant instead of being discharged to the receiving water body. The fact that these solids are returned to treatment leads to an increase in the influent solids load to the treatment stages in the liquid and solids lines.

If the design of the wastewater treatment plant takes into account the return of these solids in the overall computation of influent and effluent loads, these should be calculated by an iterative process. Three iterations are usually sufficient to accomplish the convergence of values, that is, the loads from the fourth iteration are very close to those in the third iteration. Example 2.5 clarifies the mass balance to be undertaken.

Example 2.5

For the activated sludge plant of Example 2.4, compute the mass balance for the solid loads. Assume that the supernatant and drained liquids are returned upstream of the primary clarifier. The input data are as follows:

- Influent average flow (m^3/d): 9820
- Influent SS concentration (mg/L): 379
- Secondary excess sludge load (kg/d): 1659

SS capture efficiencies:

- primary clarifier: 0.60
- thickener: 0.85
- secondary digester: 0.95
- dewatering: 0.90

VSS removal efficiency in digestion: 0.50
VSS /SS ratio in influent sludge for digestion: 0.77

Solution:

Iterative spreadsheet for the mass balance
SS loads through sludge treatment, after several iterations (kg/d).

	SS load (kg/d)					
	Iteration					
Stage	1	2	3	4	5	6
PRIMARY CLARIFIER						
Influent	3,720	4,599	4,717	4,734	4,736	4,736
Effluent	1,488	1,839	1,887	1,893	1,894	1,894
Sludge	2,232	2,759	2,830	2,840	2,841	2,842
THICKENER						
Influent	3,891	4,418	4,489	4,499	4,500	4,501
Supernatant	584	663	673	675	675	675
Thickened sludge	3,307	3,755	3,816	3,824	3,825	3,826
PRIMARY DIGESTER						
Influent	3,307	3,755	3,816	3,824	3,825	3,826
Effluent	2,034	2,310	2,347	2,352	2,353	2,353
SECONDARY DIGESTER						
Influent	2,034	2,310	2,347	2,352	2,353	2,353
Supernatant	102	115	117	118	118	118
Digested sludge	1,932	2,194	2,230	2,234	2,235	2,235
DEWATERING						
Influent	1,932	2,194	2,230	2,234	2,235	2,235
Drained	193	219	223	223	223	224
Dewatered sludge	1,739	1,975	2,007	2,011	2,011	2,012

Example 2.5 (Continued)

Notes:

- The values for the first iteration are the same as those calculated in Example 2.4.
- In the second iteration, the influent SS load to the primary clarifier is the influent load of the first iteration (3,720 kg/d), increased by the loads coming from the liquids that return to the plant (584 + 102 + 193 = 879 kg/d), leading to a total influent load of 4,599 kg/d (= 3720 + 879).
- In the second iteration, the influent load to the thickener is the sludge load from the primary clarifier, increased by the excess activated sludge load (1,659 kg/d). This secondary excess sludge load does not change from one iteration to the next. Actually, it could have been taken into account the fact that the returned solids loads also bring BOD to the system, which would imply an increased secondary excess sludge production. This analysis is beyond the scope of the present example.
- The remaining figures of the second iteration are calculated according to the same methodology used for the first iteration, as shown in Example 2.4.
- The succeeding iterations are done according to the same routine of the second iteration.
- It should be noticed that the values in the third iteration are very close to those of the last iteration, showing that the iterative process could be interrupted in the third iteration without any considerable error.
- The values of the sixth iteration are equal to those of the fifth iteration, indicating that the iterative process can be considered complete.
- It can be seen that the returned loads bring a substantial impact to the mass balance, since the values of the last iteration are higher than those from the first iteration.
- The concentrations and flows can be computed following the procedures presented in Example 2.4.

3

Main contaminants in sludge

S.M.C.P. da Silva, F. Fernandes, V.T. Soccol, D.M. Morita

3.1 INTRODUCTION

Some constituents of the wastewater, while passing through the treatment system, may increase their concentration in the sludge. Although several organic and mineral constituents in the sludge may have fertilising characteristics, others may not be desirable, due to the associated sanitary and environmental risks. These undesirable constituents can generally be grouped into:

- metals
- trace organic contaminants
- pathogenic organisms

Their presence in the sludge is extremely variable depending upon both the raw wastewater characteristics and the treatment system. Wastewaters from healthy populations present substantially less pathogens than those from unhealthy ones. In a similar way, domestic wastewater sludge has low heavy metals content, usually presenting no environmental hazard. Most chemical contaminants in the sludge are a consequence of the discharge of industrial effluents into the sewerage system.

ISBN: 1 84339 166 X. Published by IWA Publishing, London, UK.

A sound sludge management practice needs to take into account this aspect, which is often disregarded by many water and sanitation companies. A sustainable environmental policy aiming at sludge recycling requires the best economically achievable sludge quality. Water and sanitation companies must have a clear, well-defined and technically based policy of acceptance of non-domestic effluents, which avoids contaminants that could jeopardise the sludge quality and bring about the need of an expensive wastewater treatment.

Agricultural use of wastewater sludge is an acceptable practice when harmful effects can be avoided to soil, agricultural products, human health and the environment. As far as pathogenic organism contamination is concerned, a number of sludge disinfection techniques can be applied in order to reduce the pathogen densities to levels that are acceptable for agricultural use (see Chapter 6). Regarding metals and organic pollutants, there are no economically feasible techniques for their removal from sludge, especially from a developing country's perspective. Prevention is then the best strategy, because when the sludge is already contaminated, even if processed by incineration, environmental hazards may result.

3.2 METALS

3.2.1 Sources of metals in the sludge

Although metals may eventually be poisonous to plants and animals, even in the low concentrations in which they normally occur in domestic wastewaters, chronic toxicity due to their disposal is usually not reported. On the other hand, the same is not true regarding the disposal of industrial wastewaters, and mainly their sludge, because they are the major sources of concentrated metals.

Metals in wastewater are mainly due to industrial wastewater discharges from the following industries into public sewerage systems:

- electroplating
- chemical industries (organic compounds manufacturing, tanning, pharmaceutical industries)
- metal processing industries (foundries)
- chemical industries (inorganic compounds manufacturing, laundries, oil industry, dyes and pigments manufacturing)

The source of important metals found in sludges from sewerage systems that receive industrial effluents are presented in Table 3.1.

3.2.2 Potential removal of metals in biological wastewater treatment processes

The characteristics of the liquid medium define the forms in which each constituent will be present. For instance, the more alkaline is the medium, the more insoluble

Table 3.1. Main sources of metals found in sludges

Metal	Main industrial sources of contamination
Cadmium	Non-metallic mineral products: glass, cement and concrete products; metallurgical products: iron, steel, electroplating; casting works; mechanical products: electrical and electronic components; lumber industry: furniture; rubber; chemical industry: phthalic anhydride, acetylation of cellulose, benzene carboxylation, phenol/formaldehyde and aniline/formaldehyde condensation and polymerisation, inorganic compounds and elements, dyestuffs and pigments, paints and varnishes, soaps and detergents; pharmaceutical products; textile industry; photographic equipment and plastics.
Copper	Non-metallic mineral products: glass, cement and concrete products; metallurgical products: electroplating, non-ferrous metals and castings; mechanical products; electrical and electronic components; lumber industry; furniture; leather, furs and similar products; chemical industry: (a) direct chlorination of benzene, toluene, 1,4-dichlorobenzene, nitrobenzene, phthalic anhydride, methane, ethylene, propylene, etc.; (b) cellulose/acetic anhydride acetylation; (c) cracking of liquid petroleum gas, naphta/oil gas, naphta/liquid petroleum gas; (d) extraction/distillation of pyrolysis-gasoline; inorganic compounds and elements; adhesives; oil industry; plastics, plastic material products; paints and varnishes; soaps and detergents; cosmetics and fragrances; textiles; hospitals; laundries; hot water piping.
Zinc	Non-metallic mineral products: glass, cement and concrete products; metallurgical products: iron, steel, electroplating, non-ferrous metals and castings; mechanical products; electrical and electronic components; furniture; rubber; leather, furs and similar products; several chemical industries; adhesives; explosives; oil industry; oils and waxes; pesticides; plastics; plastic material products; paints and varnishes; soaps and detergents; pharmaceutical products; cosmetics and fragrances; textiles; hospitals; laundries; photographic equipment.
Nickel	Non-metallic mineral products: glass, cement and concrete products; metallurgical products: iron, steel, electroplating, non-ferrous metals; mechanical products; electrical and electronic components; furniture; leather, furs and similar products; several chemical industries; dyestuffs and pigments; explosives; plastics; plastic material products; paints and varnishes; soaps and detergents; pharmaceutical products; cosmetics and fragrances; textiles; laundries; photographic equipment.
Mercury	Metallurgical products: electroplating, non-ferrous metals; electrical and electronic components; pharmaceutical products, fungicides; electric and electronic devices; furniture; paper and cardboard; several chemical industries; adhesives; explosives; fertilisers; pesticides; plastics; plastic material products; paints and varnishes; pharmaceutical products; textile; hospitals; laboratories; photographic equipment.
Chromium	Non-metallic mineral products: glass, cement and concrete products; metallurgical products: iron and steel, electroplating, non-ferrous metals and castings; mechanical products; electrical and electronic components; lumber industry; furniture; leather, furs and similar products; several chemical industries; adhesives; dyestuffs and pigments; fertilisers; oil industry; oils and waxes; plastics; plastic material products; paints and varnish; soaps and detergents; pharmaceutical products; cosmetics and fragrances; textile; photographic equipment.

(*Continued*)

Table 3.1 (*Continued*)

Metal	Main industrial sources of contamination
Lead	Non-metallic mineral products: glass, cement and concrete products; metallurgical products: iron and steel, electroplating, non-ferrous metals; mechanical products; electrical and electronic components; furniture; rubber; leather, furs and similar products; several chemical industries; adhesives, dyestuffs and pigments; explosives; oil industry; oils and waxes; plastics; plastic material products; paints and varnish; soaps and detergents; pharmaceutical products; cosmetics and fragrances; textile; hospitals; laundries; photographic equipment; storm drainage piping and building plumbing.
Arsenic	Metallurgical products: non-ferrous metals; electrical and electronic components; lumber industry; furniture; several chemical industries; oil industry; oils and waxes; pesticides; plastics; paints and varnishes; pharmaceutical products; textile; hospitals; laboratories; laundries.
Selenium	Non-metallic mineral products: glass, cement and concrete products; metallurgical products: iron and steel, non-ferrous metals; electrical and electronic components; furniture; rubber; several chemical industries; dyestuffs and pigments; paints and varnishes; textile; photographic equipment.

Source: ADEME (1998), Morita (1993), Fernandes and Silva (1999)

lead compounds will be formed, decreasing the lead concentration in the liquid effluent. Thus, the more alkaline the medium the higher will be the lead concentration in the sludge. Metallic compounds behave similarly to lead. Hence, depending on how the treatment plant is operated, metals can be routed to the solid or liquid phase.

Furthermore, the presence of other metals and cyanide may have a synergistic or antagonistic effect. An example is the increased toxicity of copper when cyanides are present. On the other hand, in the presence of chelating agents, such as EDTA-4 and HEDTA-3, the toxicity of bivalent metals may be reduced through a complexation process. If sulphates are present, metallic sulphates can precipitate and toxicity is reduced.

Ranges of metals removal efficiencies in several wastewater treatment systems are presented in Table 3.2. The wide ranges, reflecting large variabilities and site specificity, should be noted.

The concentration of metals in the sludge is highly variable from place to place, considering all the different influencing factors. Table 3.3 shows data from some wastewater treatment plants in Brazil. As said previously, the quality of the treatment plant effluent depends upon the quality of the influent. Therefore, liquid effluents and biosolids can only be conveniently disposed of if the influent to the plant is properly characterised and checked against pre-established pollutant limits. These limits depend upon both the final disposal methods and the treatment processes, since in many cases there are concentration limitations inherent to the process.

Table 3.2. Metals removal efficiencies through several biological wastewater treatment systems

Pollutant	Treatment process	% removal	Influent concentration (µg/L)	Effluent concentration (µg/L)	References
Arsenic	Activated sludge	20–98	–	b.d.t.–160	E.P.A (1980)
	Aerated lagoon	99	–	Nd–20	E.P.A (1980)
Cadmium	Primary	7	–	–	Helou (2000)
	Trickling filter	28	25+/_23	18+/_14	Hannah *et al.* (1986)
	Activated sludge	24	25+/_23	19+/_17	Hannah *et al.* (1986)
	Aerated lagoon	–	25+/_23	–	Hannah *et al.* (1986)
	Facultative pond	32	25+/_23	17+/_9	Hannah *et al.* (1986)
	Activated sludge	0–99	–	b.d.t.–13	E.P.A (1980)
	Aerated lagoon	97	–	2	E.P.A (1980)
Lead	Primary	20	–	–	Helou (2000)
	Trickling filter (M)	48	165+/_168	86+/_79	Hannah *et al.* (1986)
	Activated sludge (M)	6.5	165+/_168	58+/_75	Hannah *et al.* (1986)
	Aerated lagoon (M)	58	165+/_168	70+/_76	Hannah *et al.* (1986)
	Facultative pond (M)	50	165+/_168	82+/_76	Hannah *et al.* (1986)
	Activated sludge	10–99		Nd–120	E.P.A (1980)
	Aerated lagoon	80–99		Nd–80	E.P.A (1980)
Copper	Primary	18	–	–	Helou (2000)
	Trickling filter (M)	60	345+/_119	137+/_77	Hannah *et al.* (1986)
	Activated sludge (M)	82	345+/_119	61+/_40	Hannah *et al.* (1986)
	Aerated lagoon (M)	74	345+/_119	89+/_61	Hannah *et al.* (1986)
	Facultative pond (M)	79	345+/_119	71+/_46	Hannah *et al.* (1986)
	Activated sludge	2–99	–	b.d.t.–170	E.P.A (1980)
	Aerated lagoon	26–94	–	7–110	E.P.A (1980)
Chromium	Primary	16	–	–	Helou (2000)
	Trickling filter (M)	52	221+/_88	107+/_130	Hannah *et al.* (1986)
	Activated sludge (M)	82	221+/_88	40+/_18	Hannah *et al.* (1986)
	Aerated lagoon (M)	71	221+/_88	65+/_106	Hannah *et al.* (1986)
	Facultative pond (M)	79	221+/_88	46+/_34	Hannah *et al.* (1986)
	Activated sludge	5–98	–	b.d.t.–2000	E.P.A (1980)
Mercury	Primary	22	–	–	Helou (2000)
	Activated sludge	33–94	–	Nd–0.9	E.P.A (1980)
	Aerated lagoon	99	–	0.1–1.6	E.P.A (1980)
Nickel	Primary	6	–	–	Helou (2000)
	Trickling filter (M)	30	141+/_93	98+/_68	Hannah *et al.* (1986)
	Activated sludge (M)	43	141+/_93	61+/_45	Hannah *et al.* (1986)
	Aerated lagoon (M)	35	141+/_93	91+/_50	Hannah *et al.* (1986)
	Facultative pond (M)	43	141+/_93	81+/_59	Hannah *et al.* (1986)
	Activated sludge	0–99	–	Nd–400	E.P.A (1980)
	Aerated lagoon	0–50	–	5–40	E.P.A (1980)
Selenium	Aerated lagoon	50–99	–	Nd–200	E.P.A (1980)
Zinc	Primary	26	–	–	Helou (2000)
	Activated sludge	0–92	–	b.d.t.–38000	E.P.A (1980)
	Aerated lagoon	34–99	–	49–510	E.P.A (1980)

b.d.t. – below detection threshold Nd – not detected M – municipal wastewater
Source: Morita (1993). Details of references: See Morita (1993).

Table 3.3. Metal contents in the sludges from some Brazilian wastewater treatment plants and restrictions to further use

	Concentration in mg/kg, dry basis							
							Maximum allowed concentration	
Chemicals	Franca (CAS)	Barueri (CAS)	Suzano (CAS)	Northern Brasília (CAS)	Belem PR (EAAS)	Londrina (UASB)	USEPA	Paraná
Arsenic	<0.006	5–68	33–202		–	–	75	–
Cadmium	0.06	8–20	2–7	<20	Nd	0.01	85	20
Lead	3	101–152	187–273	50	123	101	840	750
Copper	6	485–664	803–841	186	439	282	4,300	1,000
Mercury	4	0–1.6	15	4	1	–	57	16
Molybdenum	0.02	5–12	11	–	–	–	75	–
Nickel	0.38	211–411	269–390	2.5–5.2	73	29	420	300
Selenium	<0.06	Nd–1.4	Nd	–	–	–	100	–
Zinc	4.4	1,800–2,127	1,793–2,846	280–1,500	824	1,041	7,500	2,500

CAS: conventional activated sludge EAAS: extended aeration activated sludge UASB: upflow anaerobic sludge blanket reactor
USEPA: 40 CFR Part 503 Paraná: State of Paraná, South Brazil Nd: not detected
Source: adapted from Sapia (2000), Fernandes and Silva (1999), Helou (2000)

The limits in Table 3.3 refer mainly to the prevention of microorganisms' growth inhibition or toxicity. Therefore, one must set, for a certain constituent, the admissible load, taking into account process inhibition, effluent quality and biosolids beneficial use.

The discharge of a certain wastewater into the public sewerage system may have a variable impact on the wastewater treatment plant, depending upon dilution factors, content and type of pollutants, and the particular wastewater treatment system under operation. Proper assessment of the impact on the treatment processes may be approximated using laboratory or mathematical simulations. As a result, decisions can be taken regarding the acceptance or non-acceptance of industrial effluents to the treatment plant, taking into account inhibition of the biological treatment processes and compliance of the effluent and biosolids to pertinent legislation. The undesirable constituents are better controlled in the sources (industries and non-domestic activities).

The main purpose of the following example is to emphasise that the metals content in the wastewater has a large impact in the sludge produced. As no dilution or specific parameters from the plant were considered, the figures from the example may not be generalised.

Example 3.1

A conventional activated sludge treatment plant has an influent wastewater with 0.2 mg/L of Cd and 0.01 mg/L of Hg. Estimate the metals concentrations in the sludge, using the flow and sludge production data from Example 2.4.

Data:

- Treatment system: conventional activated sludge
- Flow: 9,820 m^3/d
- Dewatered sludge production (dry basis): 1,739 kgSS/d
- Metals concentrations: Cd = 0.2 mg/L and Hg = 0.01 mg/L

Solution:

According to Table 3.2, the following removal efficiencies may be adopted: 24% Cd and 60% Hg.

As metals removed from the liquid phase through biological, chemical and physical mechanisms are concentrated in the sludge, a simple mass balance may be computed in order to estimate the resulting concentrations. Hence:

(a) Load of metals in the influent wastewater

Cd = 9,820,000 L/day × 0.2 mg/L = 1,964,000 mg/d = 1.964 kg/d
Hg = 9,820,000 L/day × 0.01 mg/L = 98,200 mg/d = 0.098 kg/d

(b) Load of metals retained in the sludge

Cd = 1,964,000 mg/day × 0.24 = 471,360 mg/d = 0.471 kg/d
Hg = 98,200 mg/day × 0.60 = 58,920 mg/d = 0.059 kg/d

***Example 3.1* (*Continued*)**

(c) Concentration of metals in the dewatered sludge (dry basis)

Since the sludge production on a dry basis is 1,739 kg/d, the resulting concentrations are:

Cd = (471,360 mg/d) / (1,739 kg/d) = **271 mg/kg**
Hg = (58,920 mg/d) / (1,739 kg/d) = **34 mg/kg**

(d) Concentration of metals in the treatment plant effluent

$$Cd = 0.20\ mg/L \times (1 - 0.24) = 0.15\ mg/L$$
$$Hg = 0.010\ mg/L \times (1 - 0.60) = 0.004\ mg/L$$

(e) Comments

The example adopts a simplified approach for didactic reasons, and cannot be generalised, because the estimation of the expected metal concentrations in the sludge can only be done based upon specific data of the wastewater treatment plant under consideration.

General comments on the results obtained are that, in this case, the Cd contents in the sludge (271 mg/kg) are higher than the limits of 85 and 20 mg/kg set by USEPA and the State of Paraná (Brazil), respectively (see Table 3.3). Regarding Hg, the resulting concentration of 34 mg/kg complies with the USEPA standard (57 mg/kg), but not with the State of Paraná standard (16 mg/kg). Therefore, the sludge may be considered unsuitable for agricultural reuse.

It may be noticed that some metals tend to concentrate more than others in sludge (Cd = 24% and Hg = 60%). However, in both cases, even with low concentrations in the influent wastewater (Cd = 0.2 and Hg = 0.01 mg/L), they are present in substantial concentrations in the sludge dry mass (Cd = 271 mg/kg and Hg = 34 mg/kg).

This fact highlights a significant operational problem, related to the detection limit and the accuracy of the laboratorial analytical methods. In many cases, the metals are not detected with precision in the liquid phase, due to the low allowable concentrations stipulated for discharge into the public sewerage system.

A preventive measure would be the requirement of an efficient treatment for the removal of metals from industrial effluents that are traditionally known to generate contaminated effluents, before discharging to the public sewerage system.

Many legislations state that the producers of the wastes are responsible for their treatment and final disposal. Furthermore, relying only on the "end of pipe" approach is against the modern environmental management principles,

Example 3.1 (Continued)

which recommend segregation of toxic components at the source, and not its dissemination.

These facts, together with the complexity of existing models to forecast the impact of polluting loads on the treatment systems, plus the difficulties in monitoring and effectively controlling the pollutant levels discharged into the sewers, are strong arguments towards the need to enforce adequate pre-treatment of toxic industrial effluents prior to their admittance to the public sewerage systems.

3.3 TRACE ORGANICS

The main sources of organic compounds are: chemical industries, plastic industries, mechanical products, pharmaceutical industries, pesticide formulation, casthouses and steel industries, oil industry, laundries and lumber industries.

The most common organic pollutants in industrial effluents are: cyanide, phenol, methyl chloride, 1,1,1,-trichloroethane, toluene, ethyl benzene, trichloroethylene, tetrachloroethylene, chloroform, bis-2-ethyl-hexyl phthalate, 2,4-dimethyl phenol, naphthalene, butylbenzylphthalate, acrolein, xylene, cresol, acetophenone, methyl-sobutyl-acetone, diphenylamine, anilin and ethyl acetate.

Some guidelines concerning sludge organic contaminants are presented in Table 3.4.

A variety of organic compounds are receiving major attention as potential pollutants of soil, plants and water as a consequence of land application of the sludge. In the beginning, chlorinated hydrocarbons, pesticides and polychlorinated

Table 3.4. Guidelines for sludge organic contaminants (dry basis)

Constituent (mg/kg)	Denmark	Sweden	Germany
Toluene		5	
Linear alkylbenzenesulphonates	1,300		
Σ polycyclic aromatic hydrocarbons (PAH)	3	3 (sum of 6 specified PAHs)	
Nonylphenol (mono and diethoxylate)	10	50	
Di 2-ethylihexyl) phthalate	50		
Adsorbed organic halides			500
Polychlorinated biphenyl – PCB		0.4 (sum of 7 specified PCBs)	0.2 (for every 1 out of 6 specified PCBs)
Polychlorinated dibenzo-p-dioxin and polychlorinated dibenzofurans (ng/kg TEQ)			100

TEQ – toxicity equivalent in 2,3,7,8-tetrachlorodibenzo(p)dioxin
Source: da Silva *et al.* (2001)

biphenyls were the most studied compounds. Later researches have focused on compounds present in municipal wastewater treatment plants. Analyses made in 25 cities in the United States (Morita, 1993) indicated that several ester phthalates (diethyl, dibutyl) were present in 13% to 25% of the sludges in concentrations above 50 mg/kg. Toluene, phenol and naphthalene were also found in 11% to 25% of the sludges in levels higher than 50 mg/kg. Chlorinated methane, ethane and benzene were found in 3% to 36% of the sludges in concentrations above 1 mg/kg, although they were detected in relatively few sludges with values above 50 mg/kg. Trace organics were also investigated in 238 sludges in Michigan (Morita, 1993). The compounds detected in those sludges included acrylonitrile, chlorinated hydrocarbons, chlorinated benzenes, chlorinated phenols, styrene and hydroquinone. Compounds found in more than 25% of the sludges included 1,2 and 1,3-dichloropropane, 1,3-dichloropropene, tetrachlorethylene, 2,4-dinitrophenol, hydroquinone, phenol, pentachlorophenol and 2,4,6-trichlorophenol. In these compounds, the average concentrations were lower than 5 mg/kg, except for tetrachloroethylene (29 mg/kg). Styrene was found in 6 out of the 219 sludges, with concentrations varying from 99 to 5,858 mg/kg. Chlorobenze and chlorotoluene were present in 6 sludges, varying from 60 to 846 mg/kg. These data suggest that most of the trace organics may be present in the majority of the sludges with concentrations lower than 10 mg/kg. However, an industrial contribution of a specific organic compound may dramatically increase its concentration in the sludge.

3.4 PATHOGENIC ORGANISMS

3.4.1 Preliminary considerations

Organisms found in sludge may be saprophytes, commensals, symbionts or parasites. Only parasites are pathogenic and able to cause diseases in human beings and animals. Five groups of pathogenic organisms may be found in sludge: (a) helminths, (b) protozoa, (c) fungi, (d) viruses and (e) bacteria.

The pathogenic organisms may come from human sources, reflecting directly the health status of the population and the sanitation level in the region. They may also come from animal sources, whose droppings are eliminated through the water-borne sewerage system (e.g., dog and cat faeces), or else through vectors in sewers, mainly rodents.

Regarding the pathogens in sludge, epidemiological surveys showed that bacteria, viruses, helminth eggs and protozoan cysts pose risks to human and animal health. These risks are due to:

- high incidence of parasitism found in the population in different parts of the world
- long survival time of helminth eggs in the environment (*Ascaris* sp. eggs can survive up to seven years)
- low infecting dose (one egg or cyst may be enough to infect the host)

The amount of pathogens in the wastewater from a specific municipality varies greatly and depends on:

- socio-economic level of the population
- sanitation conditions
- geographic region
- presence of agro-industries
- type of sludge treatment

The population of pathogens in the sludge also varies according to the conditions listed above. However, their concentration is also influenced by the sludge treatment processes (see Chapters 2 and 6). Wastewater treatment concentrates most of the load of organisms initially present in the influent in the sludge. In the separation stages, the organisms attach to the settling solid particles. Therefore, the same initial population may be found, although in higher concentrations. Another factor to be considered is the percentage of pathogens that are present, but non-viable, because the treatment processes are able to denaturalise them, that is, these organisms lose their infectivity.

3.4.2 Helminth eggs and protozoan cysts in the sludge

Table 3.5 shows important parasites (eggs, larvae or cysts) that can be found in the sludge. Helminth eggs and protozoan cysts are shown together, because their main removal mechanism in wastewater treatment is the same (sedimentation).

Table 3.5. Important parasites whose eggs (helminths) or cysts (protozoa) can be found in the sludge

Group	Parasite	Host
Nematodes	*Ascaris lumbricoides*	Man
	Ascaris suum	Swine
	Ancylostoma duodenale	Man
	Necator americanus	Man
	Trichuris trichiura	Man
	Toxocara canis	Dogs, man
	Trichostrongylus axei	Bovines, equines, man
Cestodes	*Taenia solium*	Man, swine
	Taenia saginata	Man, bovines
	Hymenolepis nana	Man, arthropods
	Hymenolepis diminuta	Rodents, arthropods
	Echinococcus granulosus	Dogs, sheep, man
Protozoa	*Entamoeba histolytica*	Man
	Giardia lamblia	Man, dogs, cats
	Toxoplasma gondii	Cats, man, mammals, birds
	Balantidium coli	Man, swine
	Cryptosporidium	Man, bovines

Source: Thomaz Soccol (2000)

Human beings and animals become infected through different ways. The oral route is epidemiologically the most important one, although other paths such as inhalation cannot be disregarded. Infection happens (a) directly, when ingesting or handling soil or vegetables containing viable helminth eggs or (b) indirectly, when drinking contaminated water or eating raw vegetables cultivated with biosolids containing helminth eggs or protozoan cysts.

The infective dose of helminths and protozoa is very low, and in some cases a single egg or cyst may be enough to infect the host (Table 3.6).

Table 3.6. Minimum Infective Dose (MID) for protozoan cysts and helminth eggs

Pathogenic organism	MID
Protozoan cysts	10^0-10^2
Helminth eggs	10^0-10^1

Source: WHO (1989)

3.4.3 Pathogenic bacteria in the sludge

Bacteria present in the sludge come from different sources, such as human and animal intestinal flora, soil, air and water. Although the incidence of entero-bacterial diseases transmitted by sewage sludge is low, the increase in the land applications of sludge may raise the risk. Table 3.7 lists pathogenic bacteria groups, which are of concern to human and animal health.

The transmission path of most enteric bacteria is faecal-oral via water and food. The inhalation of particles containing pathogens is also possible. This form of infection represents a higher risk for individuals directly working with sludge,

Table 3.7. Important pathogenic bacteria present in the sludge (primary settled sludge)

Organism	Disease	Reservoir (in animals)
Salmonella paratyphi A, B, C	Paratyphoid fever	Domestic and wild mammals, birds and turtles
Salmonella typhi	Typhoid fever	Mammals, domestic and wild birds
Salmonella spp	Salmonellosis	Bovines and other animals
Shigella sonnei, S. flexneri, S.boydii, S.dysenteriae	Dysentery	
Vibrio cholerae	Cholera	
Yersinia enterocolitica	Gastroenteritis	Mammals, domestic and wild birds
Campyilobacter jejuni	Gastroenteritis	Domestic animals, dogs, cats and birds
Escherichia coli	Gastroenteritis	Domestic animals
Leptospira spp	Leptospirosis	Domestic and wild mammals, rats

Source: EPA (1992), ADEME (1998)

such as treatment plant employees, transportation workers and biosolid spreaders. Farmers working in biosolids fertilised soils also represent a hazard population. Some bacteria persist in infected animals that act as reservoirs. Several factors increase the possibility of pathogen transmission through biosolids application in gardens and leaf-bearing plant crops, namely:

- persistence of pathogens in the biosolids, even after treatment;
- food-borne transmission;
- pathogens reservoir in human and animal population;
- immunologically deprived people and susceptibility of pregnant women.

Anaerobically digested sludge containing bacteria applied in agricultural land may not pose considerable risks to farmers, since the survival of these pathogens in pastures is shorter than in the soil, decreasing rapidly in the upper parts of the grass, compared to the region in the vicinity of the soil.

Although the minimum infective dose for bacteria may vary from one pathogenic organism to another, it usually ranges from 10^2 to 10^6 (EPA, 1992). *Salmonella* sp. and *Shigella* sp., as the commonest pathogenic bacteria found in domestic sewage, probably are the major infecting hazard. Oral infective doses for *Salmonella* are reported to be lower than 10^3 bacteria. It is important to emphasise that bacteria are potential sources of epidemic diseases and, as a result, they must be monitored in the various wastewater treatment plants.

3.4.4 Pathogenic viruses in the sludge

Viruses are present in different types of wastewaters and sludges proceeding from various treatment processes. Their concentration is variable and depends on the population health conditions, the type of wastewater treatment process used and the stabilisation process applied for the sludge.

Viruses affect both human beings and animals, and they may be transmitted through soil, food, water, aerosols or dust. The transmission may also take place through mucosa contact and inhalation. These indirect ways of contamination represent risks for treatment plant workers, biosolid spreaders and people handling dry or liquid sludge-derived products. People living by river banks whose soil has been fertilised with biosolids are also exposed to risks. Important viruses found in domestic sludge are listed in Table 3.8.

Virus infection usually occurs via a direct path through the mouth, aspiration or ingestion of sludge. Indirect infection may happen through ingestion of pathogen-contaminated water or food. The minimum infective dose is in the order of 10^2 viruses. It must be considered that both men and animals may be infected from other sources, and these may be much more important than the biosolids.

3.4.5 Density of pathogenic organisms in sludge

The amount of pathogens found in sludge is not steady and may vary, for instance, with time (month, year, season), sampling process and other factors. Literature

Table 3.8. Important viruses found in the sludge that may affect human health

Enteric viruses	Disease	Host
Hepatitis virus	Infectious hepatitis	man
Rotavirus	Gastroenteritis	man
Enterovirus	Meningitis, encephalitis, respiratory diseases	man
Poliovirus	Poliomyelitis	man
Coxsackievirus	Meningitis, pneumonia	man
Echovirus	Meningitis, paralysis	man
Astrovirus	Gastroenteritis	man
Calicivirus	Gastroenteritis	man
Reovirus	Gastroenteritis, respiratory infections	man

Source: ADEME (1998)

data show that in primary sludge the number of helminth eggs present may be in the range of $10^3 - 10^4$ per kg TS (total solids or dry solids) or more, while viruses may range from 10 to 10^6 per kgTS.

In Brazil, Ayres *et al.* (1994) found helminth eggs densities in the order of 40 eggs/gTS in the sludge from stabilisation ponds. Thomaz Soccol *et al.* (1997), working with aerobically digested sludge (extended aeration plant in South Brazil) noticed a variable number of helminth eggs along the year, ranging from 1 to 3 eggs per g/TS, with a reduction in viability from 40% to 83%. Another plant in South Brazil had an average density of 76 eggs/gTS. Passamani *et al.* (2000) found 12 helminth eggs/gTS in a plant in Southeast Brazil. In Sao Paulo, data published by Tsutya (2000) indicate average figures between 0.25 and 0.31 eggs/gTS in a conventional activated sludge plant. In Brasília, Luduvice (2000) reported, for activated sludge plants, 16 helminth eggs per 100 mL of sludge with TS concentrations of 5%.

Table 3.9 shows density ranges for different types of pathogens and sludges.

3.4.6 Public health implications of pathogens in the sludge

Multi-purpose handling and application of domestic sewage sludge without previous stabilisation and sanitisation treatment may cause infection in human beings and animals by pathogenic agents. Infection may occur through mouth or aspiration and may happen through direct or indirect contact.

Direct way:

- during sludge spreading in soil, individuals may directly inhale or ingest pathogen-containing particles;
- through handling or ingestion of raw vegetables grown in soil fertilised with untreated sludge;
- animals are also susceptible of being directly contaminated and thus have clinical problems or serve as living reservoirs for certain pathogenic organisms.

Table 3.9. Concentration of pathogenic organisms in primary and digested sludge

Pathogen	Type of sludge	Density of pathogens
Helminth eggs	Primary sludge	10^3-10^4/kg TS
	Digested sludge	10^2-10^3/kg TS
	Partially-dewatered sludge	10^1-10^3/kg TS
	Partially-dewatered sludge from aerobic treatment	$10^2-7.5\cdot10^4$/kg TS
	Anaerobic sludge	$6.3\cdot10^3-1.5\cdot10^4$/kg TS
Protozoan cysts	Primary sludge	$7.7\cdot10^4-3\cdot10^6$/kg TS
	Digested sludge	$3\cdot10^4-4.1\cdot10^6$/kg TS
	Dewatered sludge	$7\cdot10^1-10^2$/kg TS
Bacteria	Sludge	$10^1-8.8\cdot10^6$/kg TS
	Extended aeration sludge	10^8/kg TS
Viruses	Primary sludge	$3.8\cdot10^3-1.2\cdot10^5$/L
	Digested sludge	10^1-10^3/L
	Biological sludge	$10^1-8.8\cdot10^6$/kg TS

Source: Feix and Wiart (1998), Thomaz Soccol *et al.* (1997, 2000)

Indirect way:

- drinking water contaminated with sludge containing pathogenic organisms;
- ingestion of meat from animals previously contaminated with helminth eggs (*Tænia* eggs), giving continuity to the biological cycle of the parasite.

3.4.7 Survival of pathogenic organisms

(a) Survival of pathogens in soil

Organisms such as bacteria, viruses, helminths (eggs, larvae and adults), protozoa (cysts) can be usually found in soil coming from livestock, wild animals, contaminated rivers, soil parasites, plants, or man himself. There are also free-living organisms, which do not pose hazards to livestock or men, although they may lead to erroneous diagnosis related to pathogenic agents in the sludge incorporated into soil.

When untreated sludge is applied to the soil, the pathogenic organisms remain on the surface of the soil and plants. Their survival time varies according to:

- *Survival capability of the organism itself.*
- *Soil texture and pH.* In sandy soils, the survival time of helminth eggs is lower than in wet soils. Hence, the survival time varies from place to place, and generalisations are difficult.
- *Incidence of sunlight.* Direct sunshine on the organisms leads to desiccation and reduces their survival time.

- *Ambient temperature.* Lifetime of protozoa cysts and helminth eggs in summer is shorter than in winter. In regions where autumn is cold and spring rainy, the pathogenic organisms survive for a longer period.
- *Sludge application method.* When the sludge is directly applied onto the soil, sunshine reduces the survival time of parasites. When the sludge is incorporated into the soil, it has less exposure and pathogenic life span increases. Incorporated sludge has lower direct contact risks for men and animals. How deep pathogenic organisms may reach into the ground depends upon the soil texture, geological faults and erosion areas near the application site.
- *Water retaining capability.* Low moisture sandy soils favour longer survival times for some organisms (Ancylostomatidae), while reducing for others (bacteria).
- *Microorganisms in the soil.* Competition among microorganisms may or may not favour the survival of pathogens, altering the ecological equilibrium.

Medeiros *et al.* (1999) studied the life span of sewage sludge pathogenic organisms in agricultural land, and found *Salmonella* sp. absence 42 days after sludge application. *Enterococcus* and faecal coliforms were reduced by 2 log units after 134 days. Survival of helminth eggs reached 20% after 180 days (Thomaz Soccol *et al.*, 1997).

Table 3.10 presents a general synthesis of survival time for viruses, bacteria and parasites in the soil.

(b) Survival of pathogens on crops

Pathogens survival on crops varies with the type of organism and the plant characteristics. Again, viruses, bacteria and protozoa have shorter survival time than

Table 3.10. Survival time of pathogenic organisms in the soil

Pathogenic organism	Type of soil	Mean survival time	Maximum survival time
Viruses			
Enteroviruses	Different types	12 days	100 days
Bacteria			
Faecal coliforms	Top soil	40 days	90 days
Salmonella sp.	Sandy soil	30 days	60 days
	Soil (deep layer)	70 days	90 days
Vibrio cholerae		5 days	30 days
Protozoa			
Amoebae		10–15 days	30 days
Nematodes	Irrigated soil	Several months	2–3 years
Ascaris sp.	Soil	Several months	7–14 years
Toxocara sp.	Soil	Several months	8 months
Taenia sp.	Soil	15–30 days (dry summer)	3–15 months (winter)

Source: EPA (1992), Gaspard *et al.* (1995, 1996), ADEME (1998), Schwartzbrod *et al.* (1990), Medeiros *et al.* (1999), Thomaz Soccol *et al.* (1997)

Table 3.11. Survival time of pathogenic organisms in vegetables and roots

Organism	Type of food	Maximum survival time (days)
Viruses	Beans	4
	Crops	60
Bacteria (*Salmonella*)	Potato and vegetables	40
	Carrot	10
Protozoan cysts	Vegetables	3–15
Helminth eggs	Vegetables	27–35
	Lettuce	8–15
	Tomato	28
	Beet (leaves and root)	10–30

Source: Berron (1984), quoted by ADEME (1998)

helminth eggs, particularly eggs with thicker membranes like *Ascaris* sp. and *Tænia* sp. Survival times range from 4 to 60 days for viruses, 10 to 40 days for bacteria, not more than 15 days for protozoa and several months for helminth eggs, as shown in Table 3.11.

Of course, crops that have direct contact with the soil have higher risks of having pathogenic organisms, whereas aerial plants such as apple and orange trees have lower probability of contamination. Schwartzbrod *et al.* (1990) demonstrated that helminth eggs are able to survive from 8 to 15 days in lettuce, 28 days in tomatoes, and from 10 to 30 days in radishes.

Animals grazing in pastures after biosolids application may be contaminated with pathogens. Epidemics and adverse effects on reproduction capability may occur in pathogenic bacteria contamination cases. As for parasites, it is important to mention the case of *Tænia saginata*. If there are *Taenia saginata* eggs in the applied sludge, and these eggs are ingested by the livestock, a larva phase *(Cysticercus bovis)* will evolve and may complete its life-cycle as adult *Tænia* in man's small intestine, if the infected meat is ingested. The affected meat most likely will be refused during the carcass sanitary inspection, causing serious economic damages. Animals may indirectly become infected if fed with hay grown in a sludge-applied area.

4

Sludge stabilisation

M. Luduvice

4.1 INTRODUCTION

Sewage sludge in its natural state (raw sludge) is rich in pathogenic organisms, easily putrescible and rapidly developing unpleasant smells. Stabilisation processes were developed with the purpose of stabilising the biodegradable fraction of organic matter present in the sludge, thus reducing the risk of putrefaction as well as diminishing the concentration of pathogens. The stabilisation processes can be divided into (see Figure 4.1):

- **biological stabilisation**: specific bacteria promote the stabilisation of the biodegradable fraction of the organic matter
- **chemical stabilisation**: chemical oxidation of the organic matter accomplishes sludge stabilisation
- **thermal stabilisation**: heat stabilises the volatile fraction of sludge in hermetically sealed containers

The main focus of the present chapter will be on the most widely used approach of *biological stabilisation*.

The mesophilic **anaerobic digestion** is the main sludge stabilisation process used worldwide. Aerobic digestion of sewage sludge is less popular than anaerobic digestion, and encounters application for the stabilisation of biological excess sludge in biological nutrient removal activated sludge plants. Composting is common in municipal solid waste processing plants, and is also used by a limited

ISBN: 1 84339 166 X. Published by IWA Publishing, London, UK.

Table 4.1. Sludge stabilisation technologies and final disposal methods

Treatment process	Final disposal method or use
Aerobic/anaerobic digestion	Biosolid suitable for restricted use in agriculture as soil conditioner and organic fertiliser. Usually followed by dewatering, requires further treatment (disinfection) for unrestricted uses in agriculture
Chemical treatment (alkaline stabilisation)	Used in agriculture or as daily landfill covering
Composting	Topsoil like material suitable for nurseries, horticulture and landscaping. Uses dewatered sludge
Thermal drying (pelletisation)	Product with high solids content, substantial concentration of nitrogen and free from pathogens. Unrestricted use in agriculture

Biological stabilisation → Anaerobic digestion; Aerobic digestion

Chemical stabilisation → Addition of chemicals

Thermal stabilisation → Addition of heat

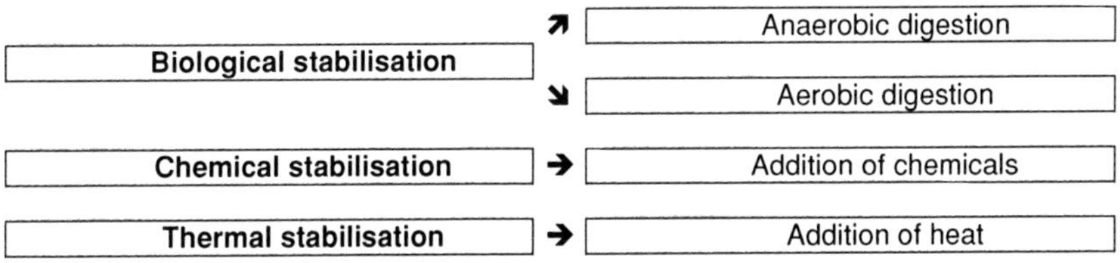

Figure 4.1. Main processes for sludge stabilisation

number of small wastewater treatment plants. Alkaline treatment and thermal drying are also processes for sludge stabilisation.

Table 4.1 shows stabilisation processes and associated sludge final disposal methods, including uses as soil conditioner or organic amendment for fields and crops.

In the various wastewater treatment systems discussed in this book and listed in Table 4.2, it is possible to notice that the degree of sludge stabilisation depends upon the wastewater treatment process adopted.

4.2 ANAEROBIC DIGESTION

4.2.1 Introduction

The word *digestion* in wastewater treatment is applied to the stabilisation of the organic matter through the action of bacteria in contact with the sludge, in conditions that are favourable for their growth and reproduction. Digestion processes may be anaerobic, aerobic or even a combination of both. Table 4.3 shows the main differences between raw sludge and digested sludge.

The anaerobic digestion process, characterised by the stabilisation of organic matter in an oxygen-free environment, has been known by sanitary engineers since the late 19th century. Due to its robustness and efficiency, it is applied to small systems such as simple septic tanks (acting as an individual solution for a

Table 4.2. Wastewater treatment processes and the corresponding degree of sludge stabilisation

	Characteristics of the sludge		
System	Primary sludge	Secondary sludge	Chemical sludge
Primary treatment (conventional)	raw		
Primary treatment (septic tanks)	stabilised		
Primary treatment with coagulation (chemically enhanced)	raw		
Facultative pond		stabilised	
Anaerobic pond + facultative pond		stabilised	
Facultative aerated lagoon		stabilised	
Complete-mix aerated lagoon + sedimentation pond		stabilised	
Facultative pond + maturation pond		stabilised	
Facultative pond + high-rate pond		stabilised	
Facultative pond + physical-chemical algae removal		non-stabilised	
Slow rate infiltration		(a)	
Rapid infiltration		(a)	
Overland flow		(a)	
Wetland		(a)	
Septic tank + anaerobic filter	stabilised	stabilised	
Septic tank + infiltration	stabilised	(a)	
UASB reactor		stabilised	
UASB + activated sludge		stabilised (b)	
UASB + submerged aerated biofilter		stabilised (b)	
UASB + anaerobic filter		stabilised	
UASB + high-rate trickling filter		stabilised (b)	
UASB + flotation		stabilised	stabilised
UASB + polishing ponds		stabilised	
UASB + overland flow		stabilised (a)	
Conventional activated sludge	raw	non-stabilised	
Extended aeration		stabilised	
Sequencing batch reactor (extended aeration)		stabilised	
Conventional activated sludge with biological N/P removal	raw	non-stabilised	
Activated sludge with chemical and biological N/P removal		non-stabilised	non-stabilised
Low-rate trickling filter	non-stabilised	non-stabilised	
High-rate trickling filter	non-stabilised	non-stabilised	
Submerged aerated biofilter		non-stabilised	
Rotating biological contactor	non-stabilised	non-stabilised	

(a): In land-disposal wastewater treatment systems, the periodic removal of formed plant biomass is necessary

(b): Assumes return of the aerobic excess sludge to the anaerobic reactor, for further thickening and digestion, together with the anaerobic sludge

Table 4.3. Comparison between raw sludge and anaerobically digested sludge

Raw sludge	Digested sludge
Unstable organic matter	Stabilised organic matter
High biodegradable fraction in organic matter	Low fraction of biodegradable organic matter
High potential for generation of odours	Low potential for generation of odours
High concentration of pathogens	Concentration of pathogens lower than in raw sludge

house) as well as in fully automated plants serving large metropolitan areas. The anaerobic digestion process underwent noticeable progresses between the First and the Second World Wars. Several concepts related to the process were improved at that time, especially in Germany, England and the United States, and are still being used today in the design of digesters.

Anaerobic digestion is a multi-stage biochemical process, capable of stabilising different types of organic matter. The process occurs in three stages:

- Enzymes break down complex organic compounds, such as cellulose, proteins and lipids, into soluble compounds, such as fatty acids, alcohol, carbon dioxide and ammonia.
- Microorganisms convert the first-stage products into acetic and propionic acid, hydrogen, carbon dioxide, besides other low-molecular weight organic acids.
- Two groups of methane-forming organisms take action: one group produces methane from carbon dioxide and hydrogen, while a second group converts the acetates into methane and bicarbonates.

4.2.2 Main requisites for sludge digestion

The efficiency and stability of the anaerobic digestion process are variables directly related to the characteristics of the raw sludge and the environment inside the digester. The raw sludge that enters the anaerobic digester is a complex mixture of materials whose characteristics are determined by the area served by the treatment plant and the wastewater treatment process adopted.

Normally, the presence of macro- and micronutrients is sufficient for ensuring the development of the anaerobic digestion process, except in the cases of digesters treating only industrial sludges. If nutrients are not a reason for concern, the presence of other materials can affect the operational performance of the sludge digester. Therefore, it is important to observe the following requisites:

Preliminary treatment. The raw sludge that comes from the primary sedimentation tanks contains, with rare exceptions, large concentrations of fibre, plastics, sand and other inert materials. These materials may pass through the preliminary treatment – screens and grit chambers – and settle with the primary sludge, causing obstruction and breakage of pipes, damage to pump rotors and to digesters mixing devices. The accumulation of sand and other materials within the digester will end up by reducing the digester net volume and, as a consequence, its efficiency. The performance of the preliminary treatment is of great importance, both to keep digestion efficiency and to reduce maintenance interventions in the digester tank.

Solids concentration. Sludge thickening is used aiming at the reduction of the volume required for digestion. Thickening is accomplished in gravity thickeners, dissolved air flotation units, or even in primary sedimentation tanks. It is desirable to have solids concentrations in the raw sludge fed to digestion in the order of 4% to 8%. Higher solids concentrations can be used, as long as the feeding and mixing units are able to handle the solids increase. Solids concentrations lower than 2.5% are not recommended, as excess water has a negative effect on the digestion process.

Inhibiting substances. Anaerobic bacteria are sensitive to several substances that, depending upon their concentrations, are capable to completely stop the digestion process. A strict control on the discharge of industrial effluents into the sewerage system and an effective legislation are the main tools to avoid the presence of toxic substances in municipal wastewater. The main inhibiting agents are hydrocarbons, organochlorinated compounds, non-biodegradable anionic detergent, oxidising agents and inorganic cations. Further details can be found in Chapter 3.

Non-biodegradable synthetic detergents are of great concern. Although their utilisation for the production of detergents has been banned in many countries, they can still be found in several other areas.

Oxidising agents like cupric ion, ferric ion and hexavalent chromium may exert an inhibiting action during the methanogenic phase of digestion, after the removal of a substantial fraction of organic matter. These ions react with sulphide ions, changing the sulphur balance inside digesters.

Inorganic cations such as sodium, potassium, calcium and magnesium, although nutrients at very low concentrations, could strongly inhibit the process at high concentrations. Optimal ammonia concentrations range from 50–1,000 mg/L; between 1,000–1,500 mg/L moderate inhibition may happen; for 3,000 mg/L and higher, strong inhibition occurs. However, these concentrations are not usual, being often associated with hog raising influents into the system.

Metals. The word *metal* in this context encompasses metals like copper, zinc, mercury, cadmium, chromium, nickel and lead. These metals can inhibit the anaerobic digestion when present individually or as metallic compounds, after reacting with enzymes needed for the process and forming insoluble complex compounds. Excluding cadmium and mercury, the other metals are considered micronutrients if present in adequate concentrations.

The destruction of organic matter during anaerobic digestion causes the metal concentration in the digested sludge to become greater than in the raw sludge (on a dry solids basis). The metal toxicity varies depending upon the metal, the presence of other metals, the pH and the concentrations of sulphide and carbonate in the sludge.

4.2.3 Process description

In a conventional activated sludge WWTP, mixed primary sludge and excess activated sludge are biologically stabilised under anaerobic conditions and converted into methane (CH_4) and carbon dioxide (CO_2). The process is accomplished in closed biological reactors known as anaerobic sludge digesters. Digester tanks are fed with sludge either continuously or in batches, and the sludge is kept inside the tank for a certain period of time previously determined during the design phase. The sludge and the solids have the same detention time in the digester.

The organic fraction of the sludge is basically made up of polysaccharides, proteins and fat. Inside the sludge digesters, colonies of anaerobic microorganisms convert the organic matter into cellular mass, methane, carbon dioxide and other micro-constituents. Inside the digester tank, three groups of mutually dependent Microorganisms coexist:

- hydrolytic acidogenic organisms
- acetogenic organisms
- methanogenic organisms

This population of microorganisms remains in a dynamic equilibrium and their concentrations vary depending upon the operational conditions within the tank.

Sulphate-reducing and denitrifying bacteria are also microorganisms occurring in anaerobic digestion and playing a fundamental role in the stabilisation process. The sulphate-reducing bacteria are responsible for the reduction of sulphate (SO_4^{2-}) to sulphide ($S^{=}$), while denitrifying bacteria reduce nitrate (NO_3^-) to gaseous nitrogen (N_2).

The redox potential inside anaerobic sludge digesters is -265 mV $\pm$ 25 mV at pH 7, and can be reduced by 60 mV per every pH unit increase. A reducing environment prevails inside the digesters. Digestion may successfully occur in pH 6–8, although pH is kept nearly neutral in practice, due to buffering capacities of bicarbonates, sulphides and ammonia. The optimum pH for anaerobic process is 7.0. Unionised acetic acid inhibits digestion in acidic pH, while unionised ammonia (NH_3) is toxic to the process in alkaline pH.

The nutritional balance within the digester is vital to control bacterial growth, and consequently, the organic matter stabilisation rate. The main nutrients, in decreasing order of importance, are nitrogen, sulphur and phosphorus. Iron, cobalt, nickel, molybdenum and selenium are major micronutrients. Iron, due to its oxidation-reduction properties and its participation in energetic metabolism, is considered the most important micronutrient in anaerobic digestion.

4.2.4 Reaction kinetics

The performance of anaerobic sludge digesters is directly linked to the concentration and diversity of the population of microorganisms present in the sludge. The solids retention time within the digester (θ_c) must be enough to ensure the maintenance of the microorganisms which have a slow growth rate, such as the methanogenic organisms, thus avoiding their wash-out from the system.

In conventional anaerobic digesters operating as complete-mix reactors, the solids retention time (sludge age) is equivalent to the hydraulic detention time, and can be determined by Equation 4.1.

$$\boxed{t = \theta_c = \frac{V}{Q}} \tag{4.1}$$

where:

t = hydraulic detention time (d)
θ_c = solids retention time (d)
V = volume of the sludge digester (m^3)
Q = influent flow to the sludge digester (m^3/d)

The slow growth rate of the methanogenic population determines the reaction time required for the anaerobic digestion process to be accomplished and, as a result, the required sludge retention time within the digester tank. Other characteristics related to θ_c and of great importance in the performance of anaerobic digesters are:

- for detention times shorter than a critical value, the process efficiency is suddenly reduced due to methanogenic organisms washout
- anaerobic digester efficiency does not increase indefinitely as detention time increases. After an optimum time is reached, the gains in efficiency are limited, not justifying further investments
- the conversion rate of the organic matter does not depend on the sludge volume fed daily to the digesters

In practice, anaerobic digesters are designed taking into consideration a detention time higher than optimum to compensate occasional operational problems such as (a) fluctuation of the sludge volume production rate, (b) inefficiency of the sludge mixing system, (c) variation of ambient temperature and (d) silting due to accumulation of inert material inside the tank.

As shown in Table 4.4, the kinetics of anaerobic digestion depends mainly on the methanogenic organisms. In normal situations, there is a perfect interaction

Table 4.4. Main characteristics of anaerobic organisms

Parameters	Acidogenic and acetogenic organisms	Methanogenic organisms
Growth rate	High	Slow
pH	Low sensitivity	High sensitivity
Temperature	Moderate sensitivity	High sensitivity
Toxic agents	Moderate sensitivity	High sensitivity
Volatile acids	Low sensitivity	High sensitivity
Redox potential	Low sensitivity	High sensitivity

between the medium and the different groups of organisms. When this balance is affected, the reaction process is also affected. For instance, the following effects of organic overloading in an anaerobic digester may be listed:

- acidogenic bacteria convert organic matter into volatile acids at a higher rate than methanogenic organisms are able to process
- volatile acids concentration is increased, reacting with alkalinity, and hence inhibiting the buffering capacity of the medium and lowering the pH value
- methanogenic organisms are inhibited due to reactor acidification
- acetogenic bacteria are inhibited due to the increasing acidification of the medium. Methane production ceases and the anaerobic digestion process starts to collapse

4.2.5 Reduction of pathogens

Raw sludge concentrates a great variety of pathogenic organisms. The concentration and type of those organisms reflect the standard of living in the treatment plant service area. The presence and concentration of certain organisms in the raw sludge may also indicate the contribution from slaughterhouses or animal related centres. This is particularly true in small wastewater treatment plants serving rural areas.

Sludge digestion significantly reduces the population of organisms, favouring the agricultural use of the sludge. Anaerobic stabilisation acts as a partial barrier between pathogenic agents and sludge users, reducing the risks of disease transmission. Chapter 6 deals with the disinfection during sludge treatment.

4.2.6 Design of anaerobic digesters

Anaerobic digesters are closed biological reactors made of concrete or steel. Inside these reactors the raw sludge is mixed – and heated, in temperate-climate countries – usually with the biogas produced, stored in floating gas holders for processing or burning. The configuration of the sludge digesters varies depending upon the area available, the need of keeping complete-mix conditions and the removal of sand and foam. Traditional anaerobic digester designs used 8–40 m diameter cylinders with 1:3 conical bottom slopes. Bottom slopes steeper than 1:3 favour sand removal but are seldom used, as they are hard to build. More recently, egg-shaped digesters have been preferred both by designers and operators, as foam and sand control are more easily accomplished thanks to its high-sloped sidewalls. Mixing requirements are not so demanding when compared with cylinder-shaped sludge digesters (Figure 4.2).

Heat loss through the walls of the anaerobic digester can be considerable, especially in cold climates. Refractory bricks on the outer wall have good aesthetics and minimise heat losses. Occasionally, half-buried sludge digesters are found, although this is not an advisable practice, since the soil, when wet, is a poor heat insulator.

Table 4.5. Typical design parameters for anaerobic sludge digesters

Parameters	Typical values
Detention time (θ_c) (d)	18–25
Volumetric organic load (kgVS/m^3·d)	0.8–1,6
Total solids volumetric load (kgSS/m^3·d)	1.0–2.0
Influent raw sludge solids concentration (%)	3–8
Volatile solids fraction in raw sludge (%)	70–80
Efficiency in total solids reduction (% TS)	30–35
Efficiency in volatile solids reduction (% VS)	40–55
Gas production (m^3/kgVS destroyed)	0.8–1.1
Calorific value of gas (MJ/m^3)	23.3
Digested sludge production (gTS/inhabitant·day)	38–50
Gas production (L/inhabitant·day)	20–30
Raw sludge heating power (MJ/kgTS)	15–25
Digested sludge heating power (MJ/kgTS)	8–15

Source: Adapted from CIWEM (1996)

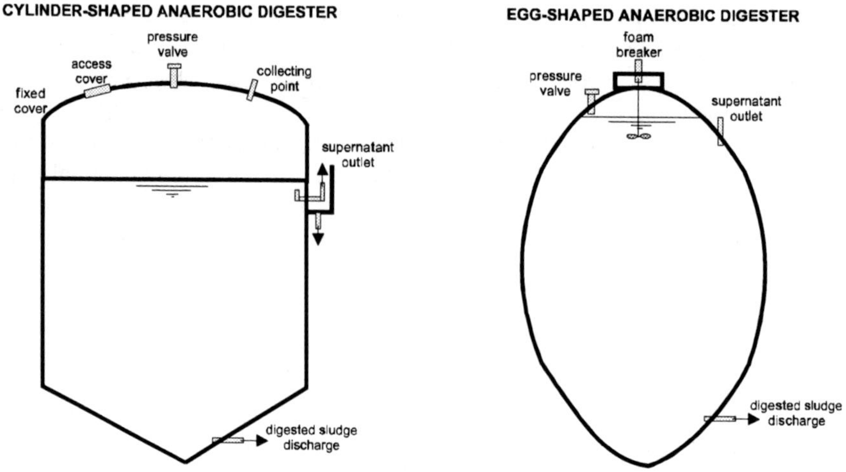

Figure 4.2. Typical formats of anaerobic digesters (adapted from WEF, 1996)

Most cylinder-shaped sludge digesters have less than 25 m diameter. Traditional design has a height-to-diameter ratio ranging from 1:2 to 1:3, and up to 33% bottom slopes. Nowadays, anaerobic digesters are also being designed with a 1:1 height:diameter ratio and a small or even zero floor slope.

Until the 1970s, the anaerobic digesters were designed for 25–30 day detention time to counterbalance possible volume losses due to sand accumulation, high water content of the raw sludge and deficiency of the mixing system. Nowadays, there is a trend to reduce the detention time to 18–25 days in warm-climate regions. Typical parameters for anaerobic sludge digesters design are listed in Table 4.5.

The required volume for the sludge digesters is given by:

$$V = \frac{\text{Influent VS load (kgVS/d)}}{\text{Volumetric organic loading (kgVS/m}^3\text{·d)}} \tag{4.2}$$

4.2.7 Mixing in anaerobic sludge digesters

As previously mentioned, the maintenance of a homogeneous sludge medium within the digester is a fundamental requirement for its good performance. Keeping homogeneity is assured through sludge mixing devices, aiming to:

- assure the internal medium uniformity from the physical, chemical and biological points of view
- quickly disperse the raw sludge when it enters the tank
- minimise thermal stratification, avoiding temperature gradients
- minimise foam formation and inert material (mainly sand) accumulation
- maximise the useful volume of the digester, minimising hydraulic short circuits and the occurrence of dead zones
- dilute the concentration of occasional inhibiting agents throughout the digester volume

The main types of sludge mixing used in anaerobic digesters are shown in Figure 4.3.

Mixing systems are either mechanical or compressed gas driven. Compressed gas systems use their own pressurised digestion gas. Gas pressurisation takes place outside the digester tank and the distribution is either through diffusers over the tank bottom or vertically along the digester sidewalls. The type of mixing system is determined by the shape and volume of the sludge digesters and the characteristics of the sludge to be digested.

Medium and large plants usually have two sludge digesters in series to optimise both the digestion process and the performance of the sludge dewatering. While

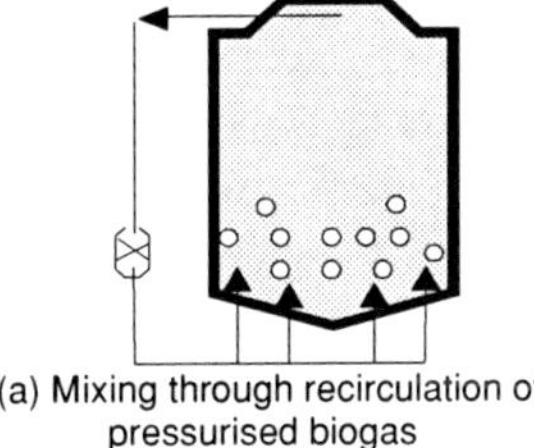

(a) Mixing through recirculation of pressurised biogas

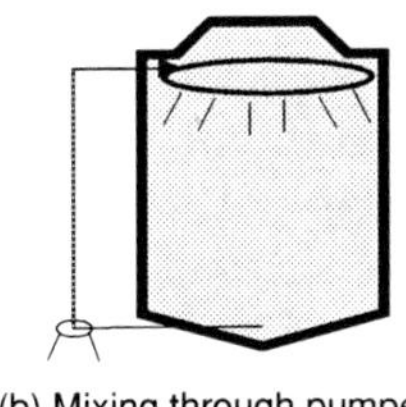

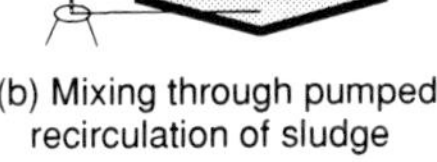

(b) Mixing through pumped recirculation of sludge

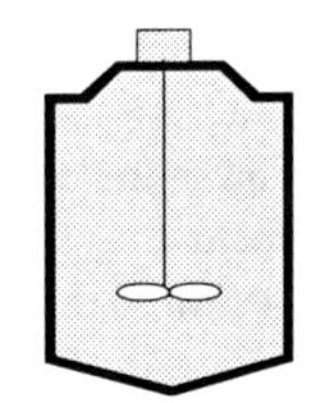

(c) Mixing through mechanical mixer

Figure 4.3. Main types of sludge mixing used in anaerobic digesters (adapted from Ferreira Neto, 1999)

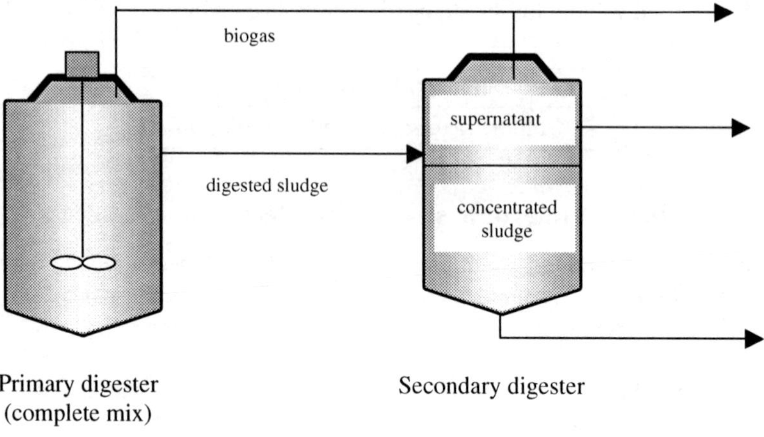

Figure 4.4. Two-stage anaerobic sludge digestion system

the *primary digester* is a complete-mix reactor responsible for fast stabilisation of the organic matter, in the *secondary digester* the separation of solid/liquid phases prevails. Secondary digesters usually do not have mixing or heating systems, except when designed to replace the primary digester during maintenance periods.

The design of secondary digesters follows the same principles presented in Table 4.5. Figure 4.4 illustrates a two-stage digestion system.

4.2.8 Biogas

Anaerobic digestion processes produce biogas, which is basically a mixture of methane (CH_4), carbon dioxide (CO_2), small concentrations of nitrogen, oxygen, hydrogen sulphide (H_2S) and traces of volatile hydrocarbons.

Biogas production in anaerobic digesters is directly associated with the raw sludge feeding. Maximum biogas production in anaerobic digesters fed at regular intervals along the day normally occurs 2 hours after each feeding.

The production rate of biogas may be estimated as 0.8 m^3/kg volatile solids destroyed, which is equivalent to approximately **25 L/inhabitant·day**. Biogas density and thermal capacity vary with the composition. The higher the methane concentration in the biogas, the higher its heating value and the lower its density. A 70%-methane biogas has a heating power of approximately 23,380 kJ/m^3 (6.5 kW/m^3). As a simple comparison, natural gas, which is a mixture of methane, propane and butane, has a heating power of 37,300 kJ/m^3 (10.4 kW/m^3).

Biogas distribution pipes must be clearly identified and kept in good working order, and confined spaces along their route in the treatment plant must be avoided. Although regularly tested for leakages and no matter how careful the maintenance staff is, it is very difficult to prevent occasional leakages. Therefore, extreme precaution is vital when using potential ignition sources, such as welding and

Poor mixture for combustion	Inflammable mixture	Mixture too rich for combustion
LEL = 5%		UEL = 15%

Figure 4.5. Combustion potential as a function of the methane concentration in the biogas/air mixture

cutting apparatus. Filament bulbs should be protected. Small exhausts should be provided for control panels in poor ventilated areas crossed by sludge pipes to avoid accumulation of gas inside the control panel, which can lead to ignition when a switch button is pushed.

Explosion may only happen when a proper combination of biogas and air occurs in the presence of a heat source (e.g., spark) with a temperature above 700 °C (ignition temperature). As biogas and air are both naturally present in the vicinity of the sludge digesters and heat sources can not be completely eliminated from the digesters supporting units (control panels, furnaces etc.), it is highly advisable to prevent biogas-air mixture situations while designing the gas piping.

The right proportion for explosion happens when methane concentration in the mixture with air reaches 5–15% (Figure 4.5). The lower explosive limit (LEL) is the minimum methane concentration (5%) needed to explode a methane/air mixture exposed to ignition. Below LEL, the methane concentration is very poor for an explosion to take place. The upper explosive limit (UEL) is 15%. Above the UEL, there is not enough oxygen to provoke an explosion.

The main characteristics of the biogas components are summarised below in terms of safety aspects:

- *Methane* (CH_4) – odourless, colourless and inflammable between 5% LEL and 15% UEL. The relative density (0.55) is lower than air, being easily dispersed. It is not toxic, although at very high concentrations may reduce the air oxygen concentrations to asphyxiating levels.
- *Carbon dioxide gas* (CO_2) – odourless, colourless and non-inflammable. The relative density (1.53) is higher than air, being asphyxiating at concentrations above 2%.
- *Hydrogen sulphide* (H_2S) – colourless, inflammable and with a characteristic rotten-egg smell. It has a relative density (1.19) nearly equal to air and 4.3% LEL and 43.5% UEL. It is irritant and asphyxiating. Concentrations higher than 1% inhibit the olfactory system and leads to unconsciousness.

The typical composition of the biogas produced in anaerobic digesters is presented in Table 4.6.

4.2.9 Temperature and heat balance

The temperature inside anaerobic digesters should be kept near 35 °C for their good operational performance. This is especially true for cold climate regions, where raw sludge temperature may be lower than 15 °C.

Table 4.6. Typical composition of biogas generated in anaerobic digesters

Gas	% (volume/volume)
Methane	62–70
Carbon dioxide	30–38
Hydrogen sulphide	50–3,000 ppm
Nitrogen	0.05–1.0
Oxygen	0.022
Hydrogen	<0.01
Water vapour	Saturation

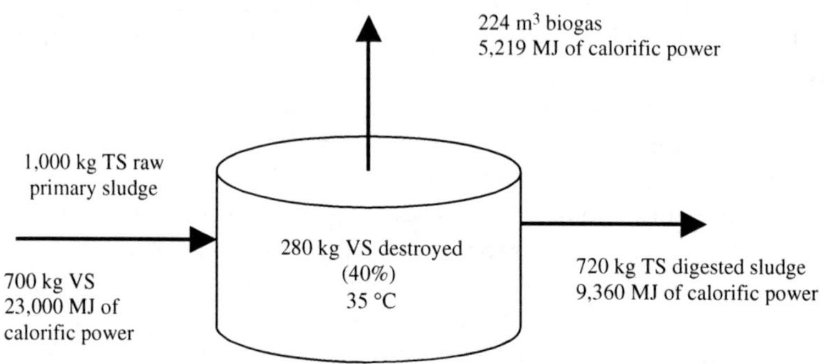

- *raw sludge heating power: 23 MJ/kgTS × 1,000 kgTS = 23,000 MJ*
- *amount of volatile solids destroyed: 700 kgTS × 0.4 = 280 kg VS*
- *amount of digested sludge: 1000 – 280 = 720 kgTS*
- *digested sludge heating power: 13 MJ/kgTS × 720 kgTS = 9,360 MJ*
- *biogas production: 0.8 m³/kg VS destroyed*
- *biogas volume produced: 280 kgVS × 0.8 = 224 m³*
- *biogas heating power: 23.3 MJ/m³ × 224 m³ = 5,219 MJ*

Figure 4.6. Example of a typical mass and heat balance during anaerobic sludge digestion

The **raw sludge** heating power ranges from **11** to **23 MJ/kgTS** on a dry-weight basis, depending upon the type of sludge and the concentration of volatile solids. The **digested sludge** has a lower heating power, which ranges from **6** to **13 MJ/kgTS** due to the smaller concentration of volatile solids.

A typical mass and heat balance within anaerobic digesters is shown in Figure 4.6.

Heating is necessary in cold weather climates to compensate for heat losses through the digesters outer surface and to raise the temperature of the raw sludge fed daily. Biogas can be used as a heat source for digester heating. Biogas is used to feed the furnace and heat the boiler, with the sludge heating indirectly accomplished by heat exchange units. In most cases, the system is self-sufficient and no further complementary external heating source is required, except during winter in very cold regions. An external heating source (e.g., fuel oil) is necessary only for the unit start-up.

The heat needed to keep anaerobic digesters near 35 °C – mesophilic digestion – is the heat needed to heat the incoming raw sludge plus the heat needed to compensate for heat losses through the digesters walls, cover and bottom. Thus:

$$Q = M_f \times C_p \times \Delta T_1 + H \tag{4.3}$$

where:

Q = sludge digester daily energy demand (kJ/d)
M_f = raw sludge mass fed to the digester (kg/d)
C_P = specific heat of water (kJ/kg·°C)
ΔT_1 = difference between the raw sludge temperature and the digester temperature (°C)
H = heat loss through the digester walls (kJ/d)

The daily heat loss through all the digester surface can be determined by:

$$H = U \times A \times \Delta T_2 \times 86.4 \tag{4.4}$$

where:

U = heat transfer coefficient (J/s·m^2·°C)
A = digester outer surface area (m^2)
ΔT_2 = difference between the digester inner temperature and the outer temperature (°C).

- *Raw sludge mass fed to digester* – $\mathbf{M_f}$: thermodynamically, a raw sludge up to 6% solids content may be considered water, with a density of 1 kg/L and specific heat (**Cp**) of 4.20 kJ/kg·°C.
- *Temperature difference* – $\mathbf{\Delta T}$: varies with the site climatic conditions. Inner digester temperature must remain between 35°C ± 3°C to assure mesophilic digestion conditions.
- *Heat transfer coefficient* – **U**: depends on the material used to build the digester tank. Literature gives U values of 2–3 J/s·m^2·°C for well-insulated digesters, whereas poorly insulated digesters may have U values of 3–5 J/s.m^2·°C.
- *Digester surface area* – **A**: includes side walls, cover and bottom area of digester tank.

Example 4.1

Design a primary anaerobic digester using data from Example 2.4.

Input data:

- Mixed sludge load to digester: 3,307 kgTS/d
- Influent sludge flow: $Q = 64.2\ m^3/d$
- VS/TS ratio = 0.77

Solution:

(a) Digester volume

Design parameters:

- Volatile solids loading rate (assumed, Table 4.5): 1.4 $kgVS/m^3 \cdot d$
- Volume reserved to the biogas in the digester: 15% of the volume needed for digestion

Volatile solids load: 3.307 kgTS/d × 0.77 kgVS/kgTS = 2,546 kgVS/d
Digesters volume (Equation 4.2): (2,546 kgVS/d)/(1.4 $kgVS/m^3 \cdot d$) = 1,819 m^3
Reserved volume for biogas accumulation: 1,819 × 0.15 = 273 m^3
Total digester volume: 1,819 + 273 = **2,092 m^3**

(b) Hydraulic detention time

Hydraulic detention time (Equation 4.1)

$$t = \theta_c = 1{,}819\ m^3/64.2\ m^3/d = \mathbf{28\ days}$$

An economic assessment of the sludge digesters construction costs may suggest higher volatile solids loading rates, which would reduce the detention time to less than 25 days.

(c) Primary digester effluent sludge (influent sludge to secondary digester)

Influent TS = 3,307 kgTS/d
Influent VS = (VS/TS) × Influent TS = 0.77 × 3,307 = 2,546 kgVS/d
Influent FS = (1 − VS/TS) × Influent TS = (1 − 0.77) × 3,307 = 761 kgFS/d

FS (fixed solids) do not change, but the VS are partially removed during digestion. According to Table 4.5, the removal efficiency of VS is between 40% and 55%. Assuming 50% (0.50) VS removal efficiency, the distribution

Example 4.1 (Continued)

of the effluent solids from the primary sludge digesters can be estimated as:

Effluent FS = Influent FS = 761 kgFS/d
Effluent VS = (1 − VS removal efficiency) × Influent VS
= (1 − 0.50) × 2,546 = 1,273 kgVS/d
Effluent TS = Effluent FS + Effluent VS = 761 + 1,273 = 2,034 kgTS/d

The sludge flow values for the primary digester effluent and influent are equal, so:

Primary effluent sludge flow = Primary influent sludge flow = **64.2 m³/d**

The TS concentration in the primary sludge digesters effluent is:

$$\text{TS conc} = \frac{\text{TS load}}{\text{Flow}} = \frac{2{,}034\ \text{kgTS/d} \times 1{,}000\ \text{g/kg}}{64.2\ \text{m}^3/\text{d}}$$

$$= 31{,}682\ \text{g/m}^3 = 31{,}682\ \text{mg/L} = \mathbf{3.17\%}$$

(d) Heat balance in digester

Raw sludge calorific power: 23 MJ/kgTS (assumed, Table 4.5)
Digested sludge calorific power: 13 MJ/kgTS (assumed, Table 4.5)
Biogas production: 0.8 m^3/kgVS destroyed (assumed, Table 4.5)
Biogas calorific power: 23.3 MJ/m^3 (assumed, Table 4.5)
Volatile solids destroyed: 1,273 kgVS/d (see Item c)
Effluent digested sludge: 2,034 kgTS/d (see Item c)

Biogas volume: 1,273 kgVS/d × 0.8 m^3/kgVS = 1,018 m^3/d
Calorific power of the raw sludge entering the digester: 3,307 kgTS/d × 23 MJ/kgTS = 76,061 MJ/d
Calorific power of biogas: 1,018 m^3/d × 23.3 MJ/m^3 = 23,719 MJ/d
Calorific power of digested sludge: 2,034 kgTS/d × 13 MJ/kgTS = 26,442 MJ/d

4.2.10 Operation and control of anaerobic sludge digesters

Operators responsible for a wastewater treatment plant know that a high level of operational performance and a peaceful ending of a daily shift may depend upon simple, easily understandable operational routines. As far as anaerobic sludge digesters operation is concerned, a good performance can be assured whenever the following factors are taken into consideration:

- suitable frequency of feeding;
- detention time higher than the methanogenic organisms growth rate;
- good operational conditions of the mixing system, assuring homogeneity inside the digester tank.

Table 4.7. Volatile acids and alkalinity ratio

Volatile acids/alkalinity ratio	Indication
<0.3	Digester is working well
0.3–0.5	Failure in digestion process
>0.8	Digestion has become acid and process collapse is imminent

Volatile acids concentration and alkalinity within the digester are closely related to each other, and the volatile acids/alkalinity ratio is a very good indicator of the quality of the digestion process. Values below 0.3 indicate good conditions within the digester, while values between 0.3–0.5 suggest deficiencies in the digestion process and call for immediate attention of the plant operator. If this ratio reaches values higher than 0.8, the digester has become acid and process collapse is imminent (Table 4.7).

Occasionally, the anaerobic digestion process may become unstable and eventually lead the digester to collapse. The instability in anaerobic digesters occurs when the series of biochemical reactions described in this chapter happens without the necessary synergy. Acid-forming bacteria outweigh the acid-consuming organisms, increasing the concentration of acids and reducing the pH in the medium. Although the causes may be varied, the instability symptoms of digestion process are common and include:

- increase of the volatile acids concentration;
- reduction of pH and alkalinity;
- reduction of methane production;
- increase of CO_2 concentration in biogas.

In such situations, the sequence of events inside the digester can be outlined as follows:

- The volatile acids/alkalinity ratio in the digested sludge reaches values above 0.3 because the volatile acids concentration has increased;
- The volatile acids start to consume alkalinity, releasing CO_2, which reduces the methane concentration and hence the biogas calorific power. The volatile acids/alkalinity ratio keeps increasing, reaching values of 0.5–0.8;
- The pH is reduced to values lower than 6.5, inhibiting methane production. The digester becomes acidified and collapses.

The collapse process described above is not immediate, taking some days to get accomplished. Therefore, it is possible to avoid it, following some measures such as:

- Through control data of the digester it is possible to determine the reasons for the process instability. A fast methanogenic process inhibition suggests the presence of highly concentrated toxic substances. A gradual inhibition indicates the presence of low-concentration toxic substances or electrical–mechanical operation problems (e.g., inoperative mixing system). A direct intervention of the maintenance staff is recommended for electrical–mechanical problems.

- The influent organic load must be verified to see if it is not above the digester capacity, since excess of organic load favours acid production and pH reduction. If the organic load is higher than recommended, feeding must be reduced until the equilibrium in the medium is reached. During digesters start-up or after maintenance periods, excessive load is a usual occurrence and acidification may happen.
- Neutral pH must be maintained through the addition of alkaline solution to the raw sludge. Should toxic substances be found in the raw sludge, dilution with non-contaminated sludge or even interruption of feeding may be carried out.
- If there is an excess of metals, sodium sulphide may be added to precipitate metallic cations. In this case, pH must be kept over 7.5, avoiding H_2S formation. The concentrations of soluble sulphides must be monitored and shall not exceed 100 mg/L.
- It may be advisable to feed anaerobic sludge from another anaerobic digester operating under stable conditions.

Feeding of the digester can be gradually brought to normal rates as soon as the digestion process shows signs of recovery.

Occasionally, the digester needs to be taken out of service for maintenance or removal of inert material deposits. The characteristics of the biogas and the amount of sludge involved require a carefully planned operation, to avoid accidents or a decrease in the plant operational performance. As previously mentioned, it is necessary to avoid an explosive biogas-air mixture inside the digester, so safety standards must be fulfilled and only skilled personnel must participate in the operation.

The following procedure can be adopted, should the anaerobic digester be taken out of service:

- stop sludge feeding
- transfer as much sludge as possible to other digesters (if existent)
- monitor gas production until it becomes negligible
- stop the mixing and heating systems
- isolate the gas outlet pipes
- if possible, complete the level of the digester with final effluent
- be sure that the methane concentration in the gas compartment is lower than 3%
- otherwise, inject nitrogen until methane concentration is reduced down to values lower than 3%
- remove the remaining mixture in the digester, taking it to dewatering or to the treatment plant headworks
- remove vent and access flanges
- start the cleaning operation

Table 4.8. Main causes of anaerobic sludge digesters collapse and corrective measures

Promoting factors of instability and consequences			Symptoms	Recommended measures
Hydraulic shock	Organic shock	Toxic load		
Excessive sludge production	Increase in sludge influent to digester	Excessive concentration of heavy metals	Increase in volatile acids concentration	Adjust alkalinity through addition of alkaline solution (e.g., lime)
Very dilute sludge in feeding	Increase of solids concentration in the influent to digester	Excessive detergent load	Alkalinity reduction pH reduction	Lower volatile acids/alkalinity ratio to <0.5
Digester silting	Change in sludge characteristics	Chlorinated organic compounds in sludge	Increase in volatile acids/alkalinity ratio	Regulate feeding routine
Excessive foam	Too fast digester start-up	Addition of oxygen	Reduction of gas production	Raise sludge concentration Restrict industrial influent to WWTP
Methanogenic organisms wash-out	Irregular feeding	Excessive sulphides	Increase in CO_2 concentration in biogas	Clean the digester Initiate a new start-up protocol

Table 4.9. Main parameters and recommended operational ranges for anaerobic digesters

Parameter	Recommended value
PH	7.0–7.2
Alkalinity (mg/L $CaCO_3$)	4,000–5,000
Volatile acids (mg/L HAc)	200

Table 4.8 summarises the main causes of anaerobic digester failure, symptoms and corrective measures.

4.2.11 Monitoring of the anaerobic digester

Sampling must be performed fortnightly (or monthly) aiming at the evaluation of the internal conditions within the digester. Under normal conditions, pH remains nearly neutral in the 7.0–7.2 range, alkalinity (as $CaCO_3$) at 4,000–5,000 mg/L, and volatile acids concentration (expressed as acetic acid) below 200 mg/L (Table 4.9). Determination of the volatile acids/alkalinity ratio, as well as data on biogas production and composition, help identify digester overloading or operational inhibitions.

Knowledge of the volatile acids composition through chromatography may also help in the digester diagnosis. Digester operation is clearly unstable if the concentration of long-chain volatile acids increase (e.g., butyric acid) compared to the concentration of short-chain volatile acids (e.g., acetic acid).

4.3 AEROBIC DIGESTION

4.3.1 Introduction

The aerobic digestion process has a great similarity with the activated sludge process. With the supply of substrate interrupted, the microorganisms are forced to consume their own energy reserves to remain alive. This is the so-called endogenous phase, where, in the absence of food supply, the biodegradable cell mass (75%–80%) is aerobically oxidised to carbon dioxide, ammonia and water. During the reaction, ammonia is oxidised to nitrate, according to the following general equation:

$$C_5H_7NO_2 + 7O_2 + \text{bacteria} \Rightarrow 5CO_2 + NO_3^- + 3H_2O + H^+ \qquad (4.5)$$

Aerobic digestion is used in activated sludge plants operating in the extended aeration mode, as well as in plants with biological nutrient removal (BNR). Sludge digestion in extended aeration processes takes place in the aeration tank, simultaneously with the oxidation of the influent organic matter process, because the food/microorganism (F/M) ratio is low. Wasted excess activated sludge in BNR processes shall not become anaerobic, otherwise, the excess phosphorus accumulated within the cell mass during the treatment process will be released as soluble orthophosphate. Under such circumstances the recommended digestion process is the aerobic digestion, which is undertaken separately, in aerobic digesters.

Currently, three types of aerobic digestion processes are used in sludge stabilisation:

- conventional aerobic digestion (mesophilic);
- aerobic digestion with pure oxygen;
- thermophilic aerobic digestion.

Differently from the anaerobic digester, the aerobic sludge digester environment is oxidant (positive redox potential). It is advisable to control the redox potential throughout the reaction process aiming to assure oxidising conditions within the digester tank. This can be achieved through a continuous potentiometer.

Aerobic sludge digesters performance depends upon the concentration of sludge and on the volume of oxygen supplied. Solids concentrations higher than 3% in conventional digesters jeopardise the oxygen transfer efficiency of the system, hampering the assimilation of oxygen by microorganisms and fostering the build-up of a reducing environment in the core of the bacterial floc. If this happens, anaerobic digestion prevails and foul odours are released. In pure oxygen digesters, solids concentration may become as high as 5%.

4.3.2 Conventional aerobic digestion

Conventional aerobic digestion stabilises the activated excess sludge in unheated open digesters through diffused air or surface mechanical aeration. The digestion occurs at a mesophilic temperature range. Sludge is usually thickened by flotation to reduce the required digestion volume. As previously mentioned, solids concentrations in the aerobic digesters should not be greater than 3%.

Aspects to be considered in the design of aerobic digesters are similar to those for activated sludge systems, such as:

- hydraulic detention time (t) which, in this case, is equal to the solids retention time, or sludge age (θ_c)
- organic loading
- oxygen demand
- power requirements (enough for supplying the oxygen demand and maintaining the sludge in suspension)
- temperature

The main design parameters for conventional aerobic sludge digesters are shown in Table 4.10.

- *Hydraulic detention time.* After 10–15 days of detention time, under a temperature around 20 °C, the concentration of volatile solids in the sludge is reduced by 40%. Higher detention time and temperature shall be provided to achieve reductions beyond 40% solids.

Table 4.10. Design parameters for conventional aerobic sludge digesters

Item	Parameter	Value
Hydraulic detention time (d) 20 °C	Excess activated sludge	10–15
	Extended aeration	12–18
	Excess activated sludge + primary sludge	15–20
Organic loading rate (kgVS/m^3·d)	–	1.6–4.8
Oxygen demand (kgO_2 /kgVS destroyed)	Endogenous respiration	~2.3
	BOD in primary sludge	1.6–1.9
Energy for keeping solids in suspension	Mechanical aerators (W/m^3)	20–40
	Diffused air (L/m^3·min)	20–40
DO in digester (mg/L)	–	1–2

VS = volatile solids
Source: Adapted from Metcalf and Eddy (1991)

- *Organic loading.* The organic loading is limited by the oxygen transfer capacity of the aeration system. Solids concentrations higher than 3% may lead to anaerobic conditions. Typical values for organic loadings are 1.6–4.8 kgVS/m^3·d.
- *Oxygen demand.* Oxygen supply must meet cell mass endogenous respiration needs and promote mixing conditions within the digester tank. The oxygen stoichiometric demand (Equation 4.5) necessary to oxidise the organic matter in sludge is 7 mols O_2/mol of cells, or approximately 2.3 kg O_2/kg of destroyed cells. The concentration of dissolved oxygen in the reactor must be kept within 1–2 mg/L. Operational data indicate that a sludge digested under such conditions is easily mechanically dewatered.
- *Mixing.* Good mixing is essential to ensure the stabilisation of the sludge in aerobic digesters. In diffused air systems, the flow for mixing is approximately 30 L air/m^3·minute, normally attained by the oxygen demand for stabilisation itself.
- *Temperature.* The solids reduction rate depends upon the temperature inside the digester: the higher the temperature, the higher is the organic matter conversion rate. Stabilisation virtually stops if temperatures fall below 10 °C. Temperature is not controlled in conventional aerobic digesters, although heat loss can be minimised in partially buried concrete tanks. Sub-surface aerators may also help to keep temperature under control.

The following parameters are utilised for assessing the operational performance of aerobic digesters:

- volatile solids reduction
- quality of supernatant
- sludge dewaterability
- odour and aspect of the digested sludge

Volatile solids reductions in aerobic digesters of 35–50% can be normally obtained with 10–15 days of detention time. If coliforms removal is a goal, the hydraulic detention time must be greater than 40 days.

Supernatant quality is a significant item in activated sludge plants designed for biological nutrient removal. Anaerobic conditions may lead to phosphorus release from the bacterial cell mass to the supernatant liquid, which is recycled back to the plant headworks, hindering the phosphorus removal effort.

Although the dewaterability of the aerobic sludge remains controversial, practical experiences have shown that it is harder to dewater than anaerobic sludge, mainly because of the destruction of the floc structure during the endogenous respiration process.

The reduction of pathogens and ammonia concentrations in the digested sludge is also a good indicator of the quality of the stabilisation process.

Example 4.2

Design an aerobic digester tank using data from Example 2.4. In the present example, only the excess activated sludge will be routed for aerobic digestion.

Input data (according to Example 2.4):

- Secondary sludge removal point: sludge recirculation line
- Excess SS load: 1,659 kgSS/d
- SS concentration in the excess sludge: 7,792 mg/L (0.78%)
- Excess sludge flow: $Q_{ex} = 213\ m^3/d$

Data from the thickened sludge (assume mechanical thickening):

- SS capture in thickener: 0.9 = 90%
- Influent SS load to digester: 1,659 kgSS/d × 0.9 = 1,493 kgSS/d
- VSS/SS ratio in excess sludge: 0.77 = 77%
- Influent VSS load to digester: 1,493 kgSS/d × 0.77 kgVSS/kgSS = 1,150 kgVSS/d
- SS concentration in thickened excess sludge: 40,000 mgSS/L = 40 kgSS/ m^3 = 4.0%
- Thickened sludge flow: $Q_{ex} = 1{,}493\ kgSS/d / 40\ kgSS/m^3 = 37.3\ m^3/d$
- Temperature: 20 °C

Design parameters:

- Digester hydraulic detention time: 15 days
- Oxygen demand: 2.3 kg O_2/kg VS destroyed
- Air density: 1.2 kg/m^3
- Oxygen concentration in the air: 23%

***Example 4.2* (*Continued*)**

Solution:

(*a*) *Volume of the aerobic digester*

$$V = 37.3\ m^3/d \times 15\ d = \mathbf{560\ m^3}$$

(*b*) *Solids loading rate*

Volatile solids loading = 1,150 kgVSS/d/560 m^3 = **2.1 kgVS/m^3·d**
(OK – within range of Table 4.10)

(*c*) *Effluent sludge from aerobic digester*

Influent TS = 1,493 kgTS/d
Influent VS = 1,150 kgVS/d
Influent FS = TS − VS = 1,493 − 1,150 = 343 kgFS/d

FS (fixed solids) remain unchanged during digestion, whereas VS are partially removed. Assuming 40% VS removal efficiency in aerobic digestion, the solids load may be computed as:

Effluent FS = Influent FS = 343 kgFS/d
Effluent VS = (1 − VS removal efficiency) × Influent VS
= (1 − 0.40) × 1,150 = 690 kgVS/d
Effluent TS = Effluent FS + Effluent VS = 343 + 690 = 1,033 kgTS/d

VS destroyed load is:

Destroyed VS load = (VS removal efficiency) × Influent VS load
= (0.40) × 1,150 = 460 kgVS/d

The effluent flow from the aerobic digester is equal to the inflow, so:

Aerobic digester effluent sludge flow = Aerobic digester influent sludge flow
= **37.3 m^3/d**

The SS concentration in the aerobic digester effluent sludge is:

$$\text{SS conc} = \frac{\text{SS load}}{\text{Flow}} = \frac{1{,}033\ \text{kgSS/d} \times 1{,}000\ \text{g/kg}}{37.3\ \text{m}^3/\text{d}}$$

$$= 27{,}694\ \text{g/m}^3 = 27{,}694\ \text{mg/L} = \mathbf{2.77\%}$$

This is the same SS concentration maintained in the aerobic digester tank. It should be noticed that this concentration is lower than 3%. Oxygen transfer to biomass is hampered for values above this limit.

Example 4.2 (Continued)

(d) Air demand

- Oxygen mass = VS load destroyed × O_2 demand = 460 kgVS destroyed × 2.3 kgO_2/kgVS = 1,058 kgO_2/d (at field conditions)
- Volume of air = (1,058 kgO_2/d)/(1.2 kgO_2/m^3 × 0.23) = 3,833 m^3/d

Air demand, assuming 10% oxygen transfer efficiency:

- Air flow needed = 3,833/0.10 = **38,330 m^3/d**
- Check air flow mixing capacity: (38,330 m^3/d)/(560 m^3) = 68 m^3 air/m^3·d = 47 L/m^3·min (OK – greater than minimum flow needed to keep solids in suspension, see Table 4.10)
- O_2 consumption at standard conditions (assuming ratio O_2 field/O_2 standard = 0.55):

$$1{,}058\ kgO_2/d/0.55 = 1{,}924\ kgO_2/d = 80\ kgO_2/hour\ (standard)$$

(e) Required power

Assuming an Oxygenation Efficiency $OE_{standard}$ = 1.6 kgO_2/kWh:

- Power = (80 kgO_2/hour)/(1.6 kgO_2/hour) = 50 kW = 68 HP

4.3.3 Aerobic digestion with pure oxygen

Aerobic digestion using pure oxygen is a variant from the conventional aerobic digestion, in which oxygen instead of air is directly supplied to the medium. The concentration of solids in the digester may be as high as 4% without any reduction in the oxygen transfer rate to the biomass.

This process is suitable for large wastewater treatment plants, where area is a prime factor, and in which pure oxygen is already being used in the biological reactor. The reaction is highly exothermic, increasing the process efficiency and favouring its use in cold-climate regions.

4.3.4 Thermophilic aerobic digestion

Section 6.4.3 also discusses the thermophilic aerobic digestion process, analysed in terms of the disinfection of the sludge.

Heat is the main by-product from the organic matter aerobic digestion process, and the temperature inside the digester can reach 60 °C, provided there is enough substrate to keep the microbiological activity.

Thermophilic aerobic digestion (TAD) started in Germany in the early 1970s aiming at the stabilisation and disinfection of sewage sludges. In the early days it was believed that thermophilic temperatures could only be reached through the use of pure oxygen. However, later experiments proved that the use of plain air should pose no problem in reaching high temperatures in the process.

Sludges from thermophilic aerobic digesters comply with class "A" biosolids rating of the US Environmental Protection Agency (USEPA), and can be unrestrictedly used in agriculture.

The process is able to stabilise about 70% of the biodegradable organic matter in the sludge after a period of only three days. To assure an autothermic reaction process, the sludge fed to the digester must have a minimum concentration of 4%, with a solids loading rate of about 50 kg TS/m^3 digester and an organic loading rate of 70 kg BOD/m^3 digester.

The main advantages of thermophilic aerobic digestion are:

- reduction of the hydraulic detention time (volume of the digester) for organic matter stabilisation;
- production of a disinfected sludge meeting USEPA biosolids rating for unrestricted reuse.

The main disadvantages of the process are:

- high capital cost;
- operational complexity;
- foam build-up on the digester surface. A freeboard of 30% of the digester height is recommended to accommodate the produced foam.

TAD's future is promising, mainly due to the increasing restraining measures for the agricultural reuse of sludge. The process still requires development, especially in terms of operational control.

4.3.5 Composting

Section 6.4.2 describes in more detail the composting process, including a design example, discussed from the perspective of pathogens removal.

Composting is an organic matter stabilisation process used by farmers and gardeners since ancient times. The composting of human faeces (night soil) is traditionally performed in China, being considered the most likely reason why fertility and structure of Chinese soil is being maintained for over 5,000 years.

The composting processes may be divided into:

- windrow composting – the simplest and most traditional composting process;
- aerated static pile composting;
- closed-reactor biological composting, or in-vessel composting.

Although versatile, sewage sludge composting demands experience and professionalism, either in the design phase, or in the operational phase.

Composting consists in the decomposition of organic matter by mesophilic and thermophilic aerobic microorganisms. Process temperatures may reach 80 °C, after which the organic matter degradation rate is reduced and the temperature quickly drops down to 60 °C. In the turned-over windrow system, the sludge is arranged in windrows with variable lengths, with the base and height varying between 4.0–4.5 m and 1.5–1.8 m, respectively. Windrows are arranged in open areas and aeration is done both through natural convection and diffusion of air, and through regular turning-over by bulldozers or equipments specifically designed for this purpose.

The main requirements for a good composting are:

- nutrients in the sludge must be balanced with a carbon:nitrogen ratio in the range of (20–30):1.
- continuous air supply should be provided to keep an oxidising environment inside the windrow. The type of material used as bulking agent is essential in this aspect. Should anaerobic digestion conditions arise within the stack, low-molecular weight volatile organic acids (propionic, butyric and acetic acids) may be generated and foul odours may be released.
- heat loss control must assure 55–65 °C for the temperature inside the windrow.
- enough moisture shall be kept within the stack. Microbiological activity is drastically reduced when moisture drops below 35–40%. However, values above 65% interfere with the aerobic digestion process, calling for sludge dewatering (>35% dry solids) prior to composting.

Due to the exothermic characteristic of the process, the heat produced within the windrow is gradually released to the atmosphere, decreasing overall moisture of the material and inactivating pathogenic organisms. To maintain a balanced stabilisation process, the windrow must be regularly turned-over, so that the material on the outer surface is incorporated within the stack.

The main bulking agents used in sewage sludge composting are urban household organic wastes and the so-called green wastes, originating from tree pruning and lawn mowing. The co-composting of these materials has the disadvantage of increasing the volume to be composted, demanding additional area availability at the wastewater treatment plant or at the solid wastes recycling plants.

Figure 4.7 shows the flowsheet of a composting process.

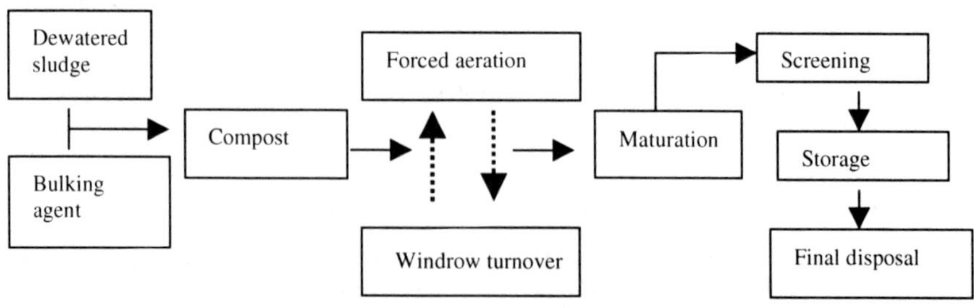

Figure 4.7. Flowsheet of a composting process

The main advantages of the composting process are:

- high-quality final product, widely accepted in farming
- possible combined use with other stabilisation processes
- low capital cost (traditional composting)

The main disadvantages are:

- need for a sludge with high-solids concentration (>35%)
- high operational costs
- need for turning-over and/or air-generation equipments
- considerable land requirements
- foul-odour generating risk

4.3.6 Wet air oxidation and incineration

Although these processes also stabilise the organic matter during the reaction process, they are discussed separately in Chapter 9.

5

Sludge thickening and dewatering

R.F. Gonçalves, M. Luduvice, M. von Sperling

5.1 THICKENING AND DEWATERING OF PRIMARY AND BIOLOGICAL SLUDGES

5.1.1 Preliminary considerations

A general description of the main processes used for thickening and dewatering were previously presented, where comparisons among the processes were made, including a balance of advantages and disadvantages. The reader is referred to these sections, which are the basis for the understanding of the present chapter.

Since the main objective of sludge thickening and dewatering is reduction of the water content in the sludge to reduce its volume, both operations are treated together in this chapter. Conditioning of the sludge, aiming at improving water removal and solids capture is also analysed. Therefore, this chapter covers the following topics and processes:

- *introductory aspects*
- *sludge thickening*
    - gravity thickeners
    - dissolved air flotation thickeners

ISBN: 1 84339 166 X. Published by IWA Publishing, London, UK.

- *sludge conditioning*
  - introductory aspects
  - organic polymers
  - inorganic chemical conditioning
- *sludge dewatering*
  - introductory aspects
  - sludge drying beds
  - centrifuges
  - filter presses
  - belt presses

An overview of all the above processes, including a general description, design criterion and operating principles is provided.

5.1.2 Water in sludge

The removal of the water content is a fundamental unit operation for the reduction of the sludge volume to be treated or disposed of. Water removal takes place in two different stages of the sludge processing phase:

- thickening
- dewatering

Sludge thickening is mainly used in primary treatment, activated sludge and trickling filter processes, having large implications on the design and operation of sludge digesters. Sludge dewatering, carried out in digested sludge, impacts sludge transportation and final disposal costs. In both cases, water removal influences sludge processing, since the mechanical behaviour of sludge depends upon its solids content.

The main reasons for sludge dewatering are:

- reduction of transportation costs to the final disposal site
- improvement in the sludge handling conditions, since the dewatered sludge is more easily conveyed
- increase in the sludge heating capacity through the reduction of the water prior to incineration
- reduction of volume aiming landfill disposal or land application
- reduction of leachate production when landfill disposal is practised

Intermolecular forces of different types are responsible for water bonding to sludge solids. Four distinct classes may be listed, according to the ease of separation:

- *free water*
- *adsorbed water*
- *capillary water*
- *cellular water*

The removal of *free water* is accomplished in a consistent way by simple gravitational action or flotation. This is what happens in gravity thickeners, where a

2% TS influent sludge leaves the unit with a solids concentration of up to 5%, leading to a sludge volume reduction of 60% or more. Another example of free water removal is the initial stage of sludge dewatering in drying beds, characterised by a rapid water loss due to percolation. *Adsorbed water* and *capillary water* demand considerably larger forces to be separated from the solids in sludge. These forces may be either chemical, when flocculants are used, or mechanical, when mechanical dewatering processes such as filter presses or centrifuges are employed. Solids contents higher than 30% may be obtained, resulting in a final product known as **cake**, with a semi-solid appearance and having a consistency compatible with spade manipulation or conveyor belt transfer. The removal of free, adsorbed and capillary water from sludge (originally at 2% TS) may result in 90–95% reduction of the original volume.

Cellular water is part of the solid phase and can only be removed through thermal forces that lead to a change in the state of aggregation of the water. Freezing and mainly evaporation are two different possibilities for cellular water separation. The thermal drying process is one of the most efficient manners for the removal of water from cakes currently available, and a 95% solids content grain-like final product can be obtained.

5.2 SLUDGE THICKENING

5.2.1 Gravity thickening

Gravity thickeners have a similar structure to sedimentation tanks. Usually they are circular in shape, centre-fed, with bottom sludge withdrawal and removal of supernatant over their perimeter. Thickened sludge is directed to the next stage (usually digestion), whereas the supernatant returns to the plant headworks. Figure 5.1 shows the schematics of a gravity thickener.

The sludge behaviour within the thickener follows the principles of zone settling and the solids flux theory. Tank sizing may be done based upon these principles or through solids and hydraulic loading rates. Table 5.1 presents typical solids loading rates as a function of the type of sludge to be thickened.

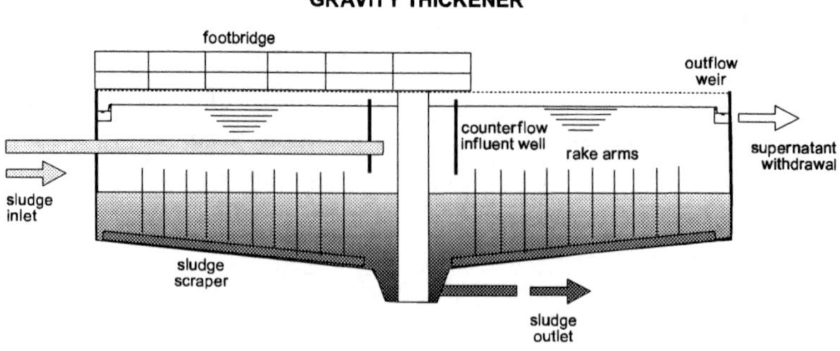

Figure 5.1. Schematic cross section of a gravity thickener

Table 5.1. Solids loading rates for the design of gravity thickeners

Source of sludge	Type of sludge	Solids loading rate ($kgTS/m^2{\cdot}d$)
Primary	–	90–150
Activated sludge	Conventional	20–30
	Extended aeration	25–40
Trickling filter	–	35–50
Mixed sludge	Primary + activated sludge	25–80
	Primary + trickling filter	<60

Sources: WEF/ASCE (1992); Jordão and Pessôa (1995); Qasim (1985)

Hydraulic loading is important in controlling excessive detention times, which could lead to the release of foul odours. Loading rates ranging from 20–30 $m^3/m^2{\cdot}d$ are therefore recommended. These values are not always achieved with the influent sludge, and final effluent recycling to the thickener is usually practised to increase the influent flow, thereby decreasing the hydraulic detention time. This flow increment is not detrimental to the thickener performance (Jordão e Pessôa, 1995).

Additional parameters according to Brazilian design standards NB-570 (ABNT, 1989) are:

- minimum sidewater height: 3.0 m
- maximum hydraulic detention time: 24 hours

Example 5.1

Design the gravity thickening unit of the conventional activated sludge system from Example 2.2.

Data:

- Population: 100,000 inhabitants
- Type of sludge: mixed (primary + activated sludge)
- Solids load in influent sludge: 7,000 kgTS/d
- Influent sludge flow: 600 m^3/d

Solution:

(a) Computation of the required surface area

From Table 5.1, the solids loading rate (SLR) may be adopted as 40 $kgTS/m^2{\cdot}d$. The required area is:

$$\text{Area} = \frac{\text{Solids load}}{\text{Solids loading rate}} = \frac{7{,}000\ \text{kgTS/d}}{40\ \text{kgTS/m}^2{\cdot}\text{d}} = 175\ \text{m}^2$$

Example 5.1 (Continued)

(b) Verification of the hydraulic loading rate

The resulting hydraulic loading rate (HLR) is:

$$\text{Hydraulic loading rate} = \frac{\text{Flow}}{\text{Area}} = \frac{600\ m^3/d}{175\ m^2} = 3.4\ m^3/m^2{\cdot}d$$

This value is lower than the range of 20–30 $m^3/m^2{\cdot}d$, recommended to avoid septic conditions in the thickener. Assuming a HLR of 20 $m^3/m^2{\cdot}d$, the following flow is needed:

$$\text{Flow} = \text{HLR} \times \text{Area} = 20 m^3/m^2{\cdot}d \times 175\ m^2 = 3{,}500\ m^3/d$$

As the available influent sludge flow is 600 m^3/d, an additional 2,900 m^3/d (= 3,500 – 600) of final effluent recycled flow is required to increase the HLR.

(c) Dimensions

Number of thickeners: n = 2 (assumed)

Area of each thickener = Total area/n = 175 m^2/2 = 87.5 m^2

Thickener diameter:

$$D = \sqrt{\frac{4.A}{\pi}} = \sqrt{\frac{4 \times 87.5\ m^2}{3.14}} = 10.6\ m$$

Sidewater depth: H = 3.0 m (assumed)

Total volume of thickeners: V = A × H = 175 m^2 × 3.0 m = 525 m^3

(d) Verification of the hydraulic retention time

The hydraulic retention time (HRT) is:

- Without final effluent recirculation: HRT = V/Q = (525 m^3)/(600 m^3/d) = 0.88 d = 21 hours (OK, less than 24 hours)
- With final effluent recirculation: HRT = V/Q = (525 m^3)/(3,500 m^3/d) = 0.15 d = 3.6 hours (OK, less than 24 hours)

5.2.2 Dissolved air flotation thickening

In the dissolved air flotation process, air is forced into a solution kept under high pressure. Under such conditions, the air remains dissolved. When depressurisation occurs, dissolved air is released, forming small bubbles, which, when rising, carry sludge particles towards the surface, from where they are skimmed off.

Table 5.2. Typical solids loading rates for dissolved air flotation thickening

Type of sludge	Solids loading rate ($kgTS/m^2 \cdot d$)	
	Without chemicals	With chemicals
Primary sludge	100–150	≤300
Activated sludge	50	≤220
Trickling filter sludge	70–100	≤270
Mixed sludge (primary + activated sludge)	70–150	≤270
Mixed sludge (primary + trickling filter)	100–150	≤300

Source: Metcalf and Eddy (1991)

Flotation thickening is widely applicable for excess activated sludge, which does not thicken satisfactorily in gravity thickeners. Dissolved air flotation is also used in treatment plants where biological phosphorus removal is practised. In these plants sludge should be kept under aerobic conditions to avoid particulate phosphorus from being released back into the liquid phase as dissolved phosphorus.

Solids loading rates used in the design of dissolved air flotation tanks are usually higher than those for gravity thickeners. Typical loading rates are shown in Table 5.2, the lower values being recommended for design purposes.

Polymers can be used in an effective way, increasing the solids capture in floated sludge. Typical dosages are between 2 and 5 kg of dry polymers per metric ton of TS.

5.3 SLUDGE CONDITIONING

5.3.1 Effects of conditioning processes

Sludge conditioning is carried out before dewatering and directly influences the processes efficiency. Conditioning may be accomplished through the utilisation of inorganic chemicals, organic chemicals or thermal treatment. The main options for conditioning and their effects on mixed sludge (primary and activated sludges) dewatering are summarised in Table 5.3.

5.3.2 Factors affecting conditioning

Conditioning aims to change the size and distribution of particles, surface charges and sludge particles interaction. The degree of hydration and the demand for chemicals and resistance to dewatering increase with the specific surface of the particles. Figure 5.2 shows relative sizes of particles from different materials.

A significant presence of colloids and thin particles with diameters normally ranging from 1μ to 10μ is very common in sewage sludges. Biomass plays a significant role in the capturing of these particles during biological treatment, diminishing sludge dewaterability and increasing the consumption of conditioning chemicals.

The main purpose of sludge conditioning is to increase particle sizes, entrapping the small particles into larger flocs. This is accomplished through coagulation

Table 5.3. Effects of conditioning processes

Item	Inorganic chemicals	Organic chemicals	Heating
Conditioning mechanism	Coagulation and flocculation	Coagulation and flocculation	Changes surface properties, splits cells, releases chemicals and causes hydrolysis
Effect on allowable solids load	Allows loading increase	Allows loading increase	Allows significant loading increase
Effect on supernatant flow	Increases solids capture	Increases solids capture	Significantly increases colour, SS, filtered BOD, $N\text{-}NH_3$ and COD
Effect on human resources	Small effect	Small effect	Requires skilled personnel and a consistent maintenance schedule
Effect on sludge mass	Significantly increases	None	Reduces existing mass, but may increase the mass through recirculation

Source: EPA (1987)

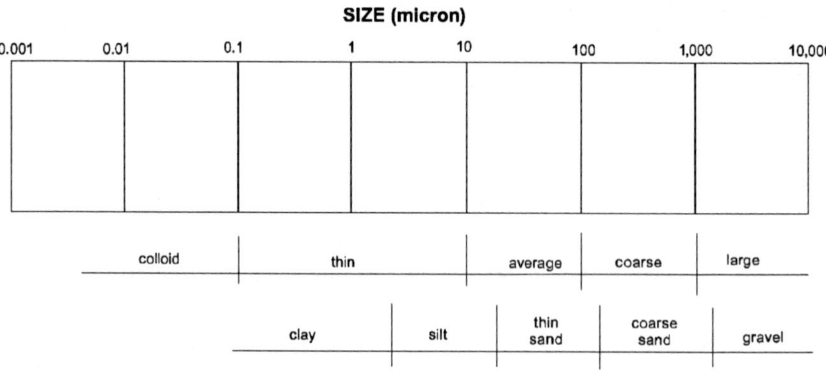

Figure 5.2. Distribution of particles size in the commonest materials (source: EPA, 1987)

followed by flocculation. Coagulation destabilises the particles, decreasing the intensity of the electrostatic repulsion forces among them. The compression of the electric double layer that surrounds each particle facilitates their mutual attraction. Flocculation allows the agglomeration of colloids and thin solids through low mixing gradients.

The amount of conditioning product to be used may vary with the sludge characteristics and the dewatering equipment adopted. The water content and the level of fine solids may change depending upon the type of sludge transportation through pipes and the storage period (weekends and longer periods). These factors affect the sludge characteristics and influence the demand for conditioners prior to dewatering.

5.3.3 Organic polymers

5.3.3.1 Main characteristics

Organic polymers are widely used in sludge conditioning. A variety of different products regarding chemical composition, performance and cost-effectiveness are available. The main advantages of organic polymers over chemical conditioners are:

- reduced sludge mass increase when compared with the mass increment when chemical conditioners are used (15–30%)
- cleaner handling operation
- reduced maintenance and operational problems
- no reduction of the calorific value of the dewatered sludge, which may be used as a fuel for incineration

Organic polymers dissolve in water to make up solutions with different viscosities. The resulting viscosity depends on their molecular weight, ionisation charge and dilution of water salt content. It is estimated that a 0.2 mg/L dosage of polymer with molecular weight of 100,000 contains around 120×10^9 active polymeric chains per litre of treated water (EPA, 1987). Polymers in solution act through attachment to the sludge particle, causing the following sequence:

- desorption of surface water
- neutralisation of charges
- agglomeration of small particulated matter through bridges among particles (bridging)

The selection of the suitable polymers should be done through routine and continuous tests involving the treatment plant operational staff and polymer suppliers. Due to changes in the characteristics of the produced sludge, tests should be carried out, whenever possible, on site, using the sludge and dewatering equipment available.

5.3.3.2 Composition and surface charges

Polymers are made up of long chains of special chemical elements, soluble in water, produced through consecutive reactions of polymerisation. They may be synthesised from individual monomers, which make up a sub-unit or a repeated unit within the molecular structure. They may also be produced through the addition of monomers or functional groups to natural polymers. Acrylamide is the most popular monomer used to produce organic synthetic polymer.

Regarding the surface charges, the polymers may be classified into *neutral* or *non-ionic, cationic* and *anionic*. Anionic flocculants with polyacrylamides introduce negative charges into the aqueous solutions, whereas cationic polyacrylamides carry positive charges. As most sludges have predominantly negative electric charges, polymers used for sludge conditioning are usually cationic. Sludge

Table 5.4. Main cationic polymers presented in dry powder (polyacrylamide copolymers)

Relative density (cationic)[1]	Molecular weight[2]	Approximate dosage (kg/mt)[3]
Low	Very high	0.25–5.00
Intermediate	High	1.00–5.00
High	Moderately high	1.00–5.00

[1] Low < 10 mole %; intermediate = 10–25 mole %; high > 25 mole %
[2] Very high = 4,000,000–8,000,000; high = 1,000,000–4,000,000; moderately high = 500,000–1,000,000
[3] mt = metric ton = 1,000 kg
Source: EPA (1987)

characteristics and dewatering equipment will determine what cationic polymer shall be more productive and cost-effective. For instance, a higher level of electric charges is needed when sludge particles are very fine, water content is high and relative surface charges are increased.

Polymers are found in powder or liquid form. Liquid polymers may be commercialised as aqueous solutions or water-in-oil emulsions. Polymers must be protected from wide temperature changes during storage, which may vary from one to several years for dry polymer powders, whereas most liquid products have storage periods from 6 to 12 months. Polymers may be found in different molecular weights and charge densities, which might greatly affect their performance in sludge conditioning.

5.3.3.3 Dry polymers

Table 5.4 shows some characteristics of dry polymers. There is a great variety of available types of polymers and a number of chemical differences influencing their performance, which are not shown in Table 5.4.

Dry polymers are available as granular powder or flocs, depending upon the manufacturing process. Due to the large quantities of chemical polymeric products, dry polymers are very active, with the concentration of active solids reaching up to 90%–95%. Dry polymers need to be stored in dry fresh places, otherwise they tend to lump and become useless.

Dry polymers require special care to be dissolved. A typical polymer feeding system is shown in Figure 5.3.

The system must include an ejector or any other kind of polymer moistening device to pre-humidify the powder being fed, which must be slowly mixed inside the tank up to complete dissolution. An extra mixing period of about 60 seconds shall be provided to ensure complete polymer dissolution. Non-dissolved polymers might cause several problems, for instance, pump and pipe clogging, scaling in filter-presses and belt-presses. The mixing period also provides time for the polymer to become effective. During this action, polymer molecules are stretched and take up

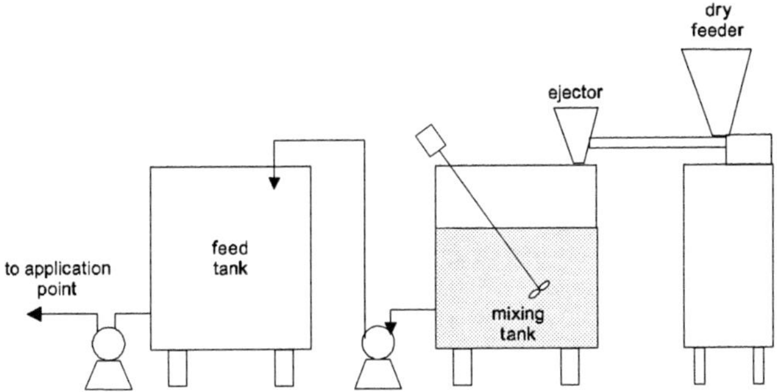

Figure 5.3. Dry polymer feeding system (EPA, 1987)

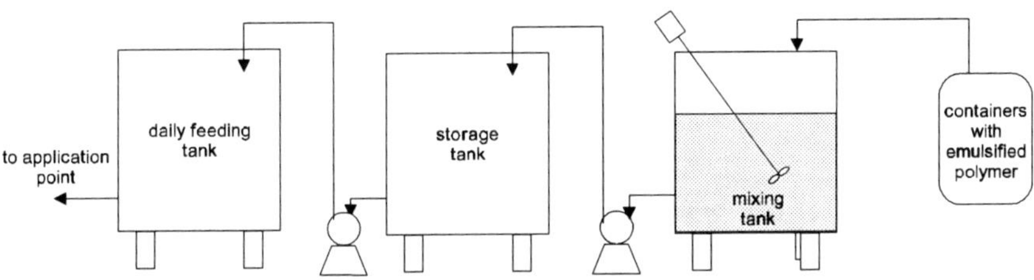

Figure 5.4. Liquid polymer feeding system (EPA, 1987)

a shape that favours sludge flocculation. If enough time is not provided, polymeric solution performance is affected.

5.3.3.4 Liquid polymers

Liquid polymers are traded with different concentrations and types of polymeric materials. They are also the active product of sludge conditioning and their dissolution depends upon the viscosity of the final solution. Stirring of concentrated polymeric solutions is not necessary, since polymers are able to form true solutions.

Liquid polymer solutions are available in 208 L containers, 1,040 L vessels, or in bulk depending on demand. Caution must be taken regarding storage in cold climates, as change in density may render its pumping impracticable.

The preparation of the dosing solution is constituted of a mixing tank and a storage tank for the diluted polymer (Figure 5.4). Normally, a 0.1% polymer solution is produced by mixing a concentrated polymer solution and water for at least 30 minutes. This solution remains stable for up to 24 hours and should be wasted after this period.

Table 5.5. Typical doses of dry polymers in different dewatering processes for several sludge types

	Belt press (kg/mt)		Centrifuges (kg/mt)	
Type of sludge	Range	Typical	Range	Typical
Raw				
– Primary	1–5	3	1–4	2
– Primary + TF	1–8	5		
– Primary + AS	1–10	4	2–8	4
– AS	1–10	5		
Anaerobic digestion				
– Primary	1–5	2	3–5	3
– Primary + AS	1–8	3	3–8	4
Aerobic digestion				
– Primary + AS	2–8	5		
Thermal conditioning				
– Primary + TF			1–3	2
– Primary + AS			3–8	4

AS = activated sludge; TF = Trickling filter sludge; mt = metric ton = 1,000 kg.
Source: Adapted from EPA (1987)

5.3.3.5 *Typical polymer dosages*

Table 5.5 shows usual polymers dosages for some mechanical dewatering processes, for various sludge types.

Example 5.2

Estimate the amount of polymer needed for conditioning the sludge from a 100,000 inhabitants conventional activated sludge treatment plant (Example 2.2). The mixed sludge undergoes anaerobic digestion before continuous centrifugation dewatering.

Solution:

(a) Amount of sludge

From Example 2.2, digested mixed sludge is: 100,000 inhabitants × 50 gTS/inhabitant· d = 5,000,000 gTS/d = 5,000 kgTS/d = 5 mt TS/day.

(b) Daily polymer consumption

From Table 5.5, the dosage should be 3–8 kg/mt. Adopting 5 kg of dry polymer/mt of TS in the sludge, the daily polymer consumption will be:

$$M_{pol} = 5\,(\text{mt TS/d}) \times 5\,(\text{kg dry polymer/mt TS in the sludge})$$

$$= 25\,\text{kg of polymer/day}$$

This value will be used for sizing the polymeric solution system and the feeding of the dewatering system.

5.3.4 Inorganic chemical conditioning

5.3.4.1 Main products

Inorganic chemical conditioning is mainly used for vacuum or pressure filtration dewatering. Mostly used chemicals are lime and ferric chloride. Although less frequently, ferrous chloride, ferrous sulphate and aluminium sulphate are also employed.

5.3.4.2 Ferric chloride

Ferric chloride is usually associated with lime for sludge conditioning, lime being added afterwards. Ferric chloride is hydrolysed in water and forms positively charged iron complexes that neutralise the negative surface charges in the sludge solids, allowing their aggregation. Ferric chloride also reacts with the sludge bicarbonate alkalinity, forming hydroxides that act as flocculants, according to the following reaction:

$$2\,FeCl_3 + 3\,Ca(HCO_3)_2 \rightarrow 2\,Fe(OH)_3 + 3\,CaCl_2 + 6\,CO_2 \qquad (5.1)$$

Ferric chloride solutions are usually employed in 30%–40% concentrations, as received from the supplier. Its dilution is not recommended as this may promote hydrolysis and precipitation of ferric hydroxide. Solutions can be stored for a long time without deterioration, although crystallisation may occur under low temperatures (below $-1\,^{\circ}C$, a 45% $FeCl_3$ solution crystallises). Ferric chloride is a very corrosive product, requiring special pumps, resistant storage materials and careful operational procedures.

5.3.4.3 Lime

Slaked lime is usually utilised together with ferric chloride mainly for pH and odour control, as well as pathogen reduction. The resulting product of the reaction of lime with bicarbonate ($CaCO_3$) yields a granular structure in the sludge, increasing its porosity and reducing its compressibility.

Lime is traded as quicklime (CaO) or as slaked lime [$Ca(OH)_2$]. Before use, quicklime must be slaked with water, producing $Ca(OH)_2$. The slaking operation releases considerable heat, demanding proper equipment and care to protect plant workers. When selecting the slaking process, the CaO contents in the different types of quicklime must be taken under consideration (Table 5.6). Slaked lime

Table 5.6. CaO contents in different types of quicklime

Rating	CaO content (% of mass)
Low content	50–75
Intermediate content	75–88
High content	88–96

Table 5.7. Dosages of conditioners for filter presses dewatering

Type of sludge	Filter press (kg/mt)	
	$FeCl_3$	CaO
Raw		
– Primary	40–60	10–140
– Activated sludge (AS)	70–100	200–250
Anaerobically digested		
– Primary + AS	40–100	110 –300
Thermally conditioned	Nil	Nil

Source: Adapted from EPA (1987), WEF (1996)

must be stored in dry places to prevent hydration reactions with air moisture, which would render it useless.

On the other hand, slaked lime requires no slaking, mixes easily with water, releasing negligible heat and does not demand special storage requirements. However, because slaked lime is more expensive and less available than quick lime, slaking of quicklime on site may be more economical in plants consuming more than 1–2 mt of lime per day.

5.3.4.4 Applied dosages

Chemical conditioning increases approximately one metric ton of sludge mass for every metric ton of lime or ferric chloride used. Chemical conditioning using lime stabilises the sludge, but reduces its heating value for incineration. Dosing ranges for filter presses and different types of sludges are presented in Table 5.7.

Example 5.3

Estimate the amount of chemicals needed for conditioning the sludge from a 100,000 inhabitants conventional activated sludge treatment plant (Example 2.2). The mixed sludge undergoes anaerobic digestion before filter press dewatering in continuous operation.

Solution:

(a) Amount of sludge

From Example 2.2, the production of digested mixed sludge is: 100,000 inhabitants $\times$ 50 gTS/inhabitant.d = 5,000,000 gTS/d = 5,000 kgTS/d

(b) Maximum quantity of $FeCl_3$ needed per day

The required quantity must be calculated incorporating a good safety margin, adopting the upper range from Table 5.7 for anaerobically digested mixed sludge. Assuming 100 kg of $FeCl_3$ per mt of TS:

***Example 5.3* (*Continued*)**

$$M_{FeCl3} = 5{,}000\ (kgTS/d) \times 100\ (kg\ FeCl_3/1{,}000\ kg\ TS)$$
$$= 500\ kg\ FeCl_3/day$$

(c) Volume of the solution with 40% $FeCl_3$

The solution of 40% $FeCl_3$ has 1.0 kg of $FeCl_3$ per 1.77 L of solution. Therefore:

$$V_{FeCl3} = 500\ (kg\ FeCl_3/day) \times 1.77\ (L/\ kg\ FeCl_3) = 885\ litres\ of\ solution/day$$

(d) Quantity of CaO needed

Using 300 kg of CaO per mt of TS (upper limit of Table 5.7), the required amount is:

$$M_{CaO} = 5{,}000\ (kgTS/d) \times 300\ (kg\ CaO/1{,}000\ kg\ TS)$$
$$= 1{,}500\ kg\ CaO/day$$

(e) Quantity of quicklime needed

Using a quicklime with 90% CaO in its composition, the daily amount is:

$$M_{quicklime} = 1{,}500\ (kg\ CaO/day) \times (1/0.9)\ (kg\ quicklime\ /\ kg\ CaO)$$
$$= 1{,}667\ kg\ quicklime/day$$

(f) Extra daily solids production due to conditioning

It is estimated that the extra production will be 1.0 kg TS/kg ($FeCl_3$ + lime) added.

$$M_{extra\ sludge} = 1.0 \times [500\ kg\ (FeCl_3/day) + 1{,}667\ (kg\ quicklime/day)]$$
$$= 2{,}167\ kg\ TS/day$$

(g) Total mass of dry solids produced daily in the treatment plant

$$M_{dry\ sludge} = 5{,}000\ (kgTS/d) + 2{,}167\ (kg\ TS\ extra\ sludge/day)$$
$$= 7{,}167\ kg\ TS/day$$

(h) Mass of sludge (wet basis) produced daily after dewatering (sludge cake at 30% TS)

$$M_{sludge\ cake} = 7{,}167\ (kg\ TS/day) \times (100\%/30\%) = 23890\ kg\ sludge/day$$

Table 5.8. Main factors influencing dewatering efficiency in mechanical processes

Factors influencing dewatering	Causes
Proportions of primary and secondary sludge in the sludge to be dewatered	Secondary sludge retains twice the amount of water held by primary sludge (in kg of water/kg TS) during dewatering
Type of secondary sludge	The longer the sludge age, the larger the amount of water kept in the sludge. Bulked sludges (with excessive filamentous organisms) retain more liquid than non-bulked ones
Sludge conditioning	The use of chemicals for sludge conditioning may substantially improve the performance of the dewatering process
Type and age of the dewatering equipment	A number of variants of the same dewatering equipment may present different efficiencies. Older equipment is usually less efficient than modern ones
Design and operation	The design and operation of the dewatering units directly influences the cake TS contents. Equipment running near their limiting capacity tend to produce wetter cakes (3% to 5% less TS). Dryer cakes are obtained with lower loading rates
Industrial discharges	Industrial discharges into the sewerage system may positively or negatively affect the performance of the dewatering stage

5.4 OVERVIEW ON THE PERFORMANCE OF THE DEWATERING PROCESSES

Table 5.8 shows important factors influencing the solids concentration in sludge cake following mechanical processes.

Table 5.9 compares the dewatering efficiencies for natural and mechanical (with conditioning) dewatering processes. However, it should be noticed that the figures presented may vary from plant to plant.

The best results for mechanical dewatering are obtained using filter presses (plate or diaphragm). This is a discontinuous process, that may produce cakes 6%–10% dryer than continuous processes. With the inclusion of the diaphragm, this difference may reach 9%–15% more TS. Ferric chloride and lime are usually the preferred inorganic conditioning agents applied in filter press and vacuum filter dewatering. Both types of dewatering equipment become slightly less efficient (2%–5% wetter sludges) if organic instead of inorganic polymers are used.

Centrifuges and belt presses come next in terms of dewatering efficiency, with similar results for different types of sludge. Belt presses with a wider range of pressure adjustment capability may produce 2%–3% dryer cakes than centrifuges.

Table 5.9. Total solids contents (% TS) for different types of sludge and dewatering processes

Sludge	Process	Total solids (% TS), scale 0–54 (shaded range)
Primary raw	Drying bed	Not applicable
	Sludge lagoon	Not applicable
	Centrifuge	29–34
	Vacuum filter	25–32
	Belt press	31–37
	Filter-press	40–46
Activated sludge	Drying bed	Not applicable
	Sludge lagoon	Not applicable
	Centrifuge	14–20
	Vacuum filter	12–18
	Belt press	13–19
	Filter-press	27–33
Trickling filter	Drying bed	Not applicable
	Sludge lagoon	Not applicable
	Centrifuge	14–20
	Vacuum filter	13–17
	Belt press	13–20
	Filter-press	27–32
Digested mixed sludge	Drying bed	30–40
	Sludge lagoon	25–35
	Centrifuge	20–25
	Vacuum filter	17–23
	Belt press	20–25
	Filter-press	29–35

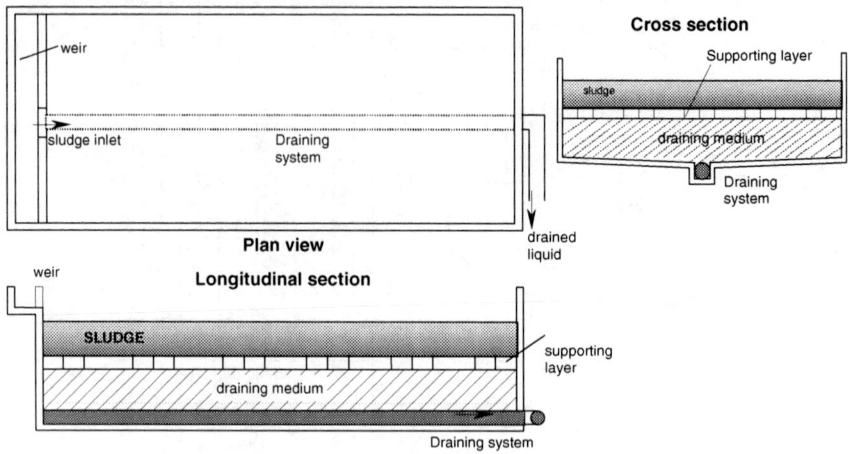

Figure 5.5. Diagram of a sludge drying bed (Gonçalves, 1999)

5.5 SLUDGE DRYING BEDS

5.5.1 Main characteristics

Water is removed by evaporation and percolation. The process consists of a tank, usually rectangular, of masonry or concrete walls and a concrete floor. Inside the tank the following elements enable the drainage of the sludge water (Figure 5.5):

- draining medium
- supporting layer
- draining system

Draining medium. Allows percolation of the liquid present in the sludge through top layers of sand and bottom layers of gravel. The layers are placed so that the grain sizes rank from top to bottom in increasing diameter, ranging from 0.3 mm in the upper part (sand) to 76 mm in the lower part (gravel) (Figure 5.6). The total depth of the layers is approximately 0.50 m.

Supporting layer. The supporting layer is built with hard burnt brick or other material able to withstand the dry-sludge removal operation. The elements are usually arranged as shown in Figure 5.7 with 20–30 mm joints with coarse-grained sand. The supporting layer allows a better distribution of the sludge, avoids clogging of the draining medium pores and ensures the dewatered sludge removal without disturbance of the draining medium layers.

Drainage system. It is made up of 100 mm pipes laid out over the tank floor, with open or perforated joints aiming to drain all the liquid percolated through the draining medium layers. The distance between the pipe drains shall not surpass

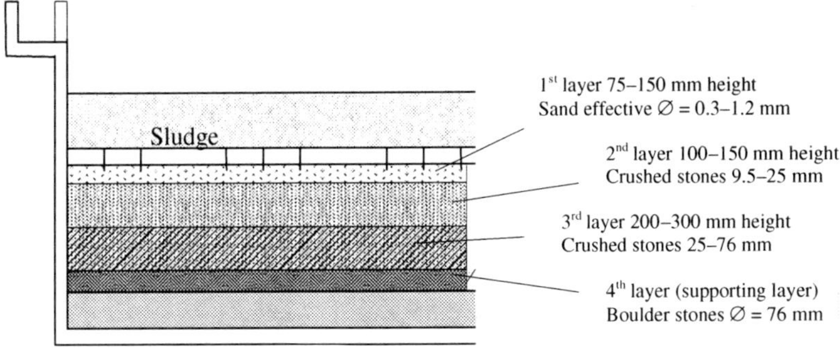

Figure 5.6. Details of the draining medium (Gonçalves, 1999)

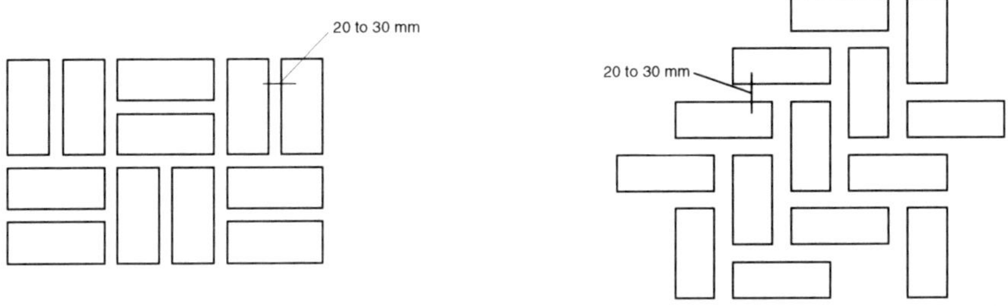

Figure 5.7. Detail of the bricks arrangement in the supporting layer (Gonçalves, 1999)

3 m. The floor of the drying bed must be even and impermeable, with minimum 1% slope towards the main draining collector.

5.5.2 General aspects of sludge dewatering in drying beds

When well digested (small fraction of biodegradable solids content), sludge subjected to natural drying has satisfactory characteristics, allowing dewatering within a short period of time (Jordão and Pessôa, 1995). An example is the bottom sludge from stabilisation ponds, which presents these characteristics and usually has reached sufficient biological stability to allow a liquid–solid separation with no need of prior treatment.

Sludge drying beds may be open-air constructions or covered for protection against rainfall. Drying is undertaken as a batch process, sequentially routing the sludge to several drying beds (van Haandel and Lettinga, 1994). According to Hess (1973), digested sludges submitted to high hydraulic pressures, either in clarifiers or in sludge digesters, may present interstitial water saturated with gases such as CO_2 or methane. This sludge may float in drying beds due to density differences between digested sludge and water. During most of the dewatering period, the water percolates easily through the draining bed, up to the moment when the sludge deposits itself and changes into a thick pasty mass. From this point on, the percolation virtually ends and drying is achieved through natural evaporation.

According to Imhoff (1966), the level of sludge stabilisation may be derived from the final characteristics of the dewatered sludge, as described below:

- dry sludge with scarce and thin cracking: indication of a well-digested sludge with a low water content
- large number of medium-sized cracks: indication of a digested sludge with high water content
- small quantities of wide cracks: indication of a poorly-digested sticky sludge, requiring long drying periods

Besides the sludge physical characteristics, climatic conditions also influence the performance of this type of process. Natural drying may promote a considerable removal of pathogenic organisms due to sunlight exposure (van Haandel and Lettinga, 1994).

When the solids content reaches around 30%, the sludge is ready to be withdrawn from the drying bed, to avoid difficulties associated with later removal. Prolonged stay of dry sludge in the drying beds leads to the growth of vegetation, indicating a poor plant management.

5.5.3 Design of drying beds

(a) Design based on loading rates

Sizing of drying beds may use empirical rates, either derived from experience on similar applications, or obtained through tests carried out under controlled conditions, specific to the focused situation. The main variables are:

- sludge production at the treatment plant
- sludge characteristics concerning total solids and volatile solids contents
- cake total solids content, which will determine the drying period
- sludge layer height on the drying bed

Brazilian standards (ABNT, 1989) recommendations are summarised as follows:

- solids loading rate: SLR $\leq$15 kg TS/m^2 of bed surface per drying cycle
- at least two drying beds should be provided
- maximum transportation distance for the removal of dried sludge within each bed: 10 m

Example 5.4 illustrates the design of sludge drying beds based on solids loading rates.

(b) Design based on the concept of productivity

Another design possibility is proposed by van Haandel and Lettinga (1994), as illustrated in Example 5.5, based upon field data and using the concept of productivity.

Example 5.4

Design a drying bed system for a 100,000 inhabitants treatment plant with UASB reactors (Example 2.1), using loading rates criteria. The drying period has been estimated to be 15 days, based upon existing drying beds performance. The dry sludge shall be removed after 5 days.

Solution:

(a) Amount of sludge to be dewatered

According to Tables 2.1 and 2.2, the per capita sludge production is 12–18 gSS/inhabitants·d, and the per capita volumetric production is 0.2–0.6 L/inhabitant·d for sludges from UASB reactors. Assuming intermediate values, the total sludge production for the 100,000 inhabitants may be computed as follows:

$$\text{SS load in the sludge: } M_s = 100{,}000 \text{ inhabitants} \times 15 \text{ g/inhabitants·d}$$
$$= 1{,}500{,}000 \text{ gSS/d} = 1{,}500 \text{ kgSS/d}$$
$$\text{Sludge flow: } Q_s = 100{,}000 \text{ inhabitants} \times 0.4 \text{ L/inhabitant·d}$$
$$= 40{,}000 \text{ L/d} = 40 \text{ m}^3\text{/d}$$

These values are equal to those computed in Example 2.1.

(b) Operational cycle time of the drying bed

$$T = T_d + T_c$$

where:

T_d = drying time (days)
T_c = cleaning time (days)

$$T = 15 + 5 = \mathbf{20\ days}$$

(a) Volume of dewatered sludge per cycle

$$V_s = Q_s \times T$$

Example 5.4 (Continued)

where:

V_s = volume of dewatered sludge per cycle (m^3)
Q_s = sludge flow (m^3/day)

$$V_s = 40\ (m^3/d) \times 20\ (d) = 800\ m^3/\text{cycle}$$

(b) Area required for the drying bed

$$A = (M_s \cdot T)/SLR = [1{,}500\ (\text{kg TS/d}) \times 20\ (d)]/15\ (\text{kg TS/m}^2) = \mathbf{2{,}200\ m^2}$$

where:

A = drying bed area (m^2)
SLR = nominal solids loading rate (adopted as **15 kg TS/m²**)

The per capita required area is:

Per capita area = 2,200 m^2/100,000 inhabitants = 0.022 m^2/inhabitant

(c) Dimensions of the drying cells

A total of **22 cells** (greater than the cycle time of 20 days) with 100 m^2 each will be used. Each cell will be **10 m wide** and **10 m long**.

(d) Height of the sludge layer after loading operation at the drying bed

The sludge is calculated by:

$$H_s = V_s/A = 800\ (m^3/\text{cycle})/2{,}200\ (m^2/\text{cycle}) = \mathbf{0.36\ m}$$

Example 5.5

An anaerobic pond treating sewage from 20,000 inhabitants has accumulated 723 m^3 of sludge after 2 years of uninterrupted operation. The removal of 80% of the accumulated sludge volume shall be accomplished, and subsequent dewatering in drying beds within the plant area is being planned. Compute the required area for the drying beds using the concept of productivity, assuming sludge is removed with 92% moisture (water) content (8% TS), and should be dewatered to reach a moisture content of 73% (27% TS).

Solution:

(a) Solids mass to be removed

$$M_{sludge} = V_{sludge} \times SC_i \times \rho_S$$

Example 5.5 (Continued)

where:

V_{sludge} = volume of sludge to be removed = 723 $m^3 \times 0.8 = 579\ m^3$
SC_i = initial sludge solids content = 8% (moisture = 92%)
ρ_S = sludge density = 1,020 kg/m^3

$$M_{sludge} = 579 \times 0.08 \times 1{,}020 = 47 \times 10^3\ kgTS$$

(b) Productivity

The productivity relates the applied solids load ($kgTS/m^2$) and the drying period (days) for a particular moisture (water content). As shown in Figure 5.8, this is the ratio of TS mass per unit area and per unit time. The curves shown in Figure 5.8 are derived from actual field data from one specific anaerobic pond in Southeast Brazil (Gonçalves, 1999).

For sludges from UASB reactors with a final water content of 70% (30% TS), productivity values of 1.65 $kgTS/m^2{\cdot}d$ were gathered in a warm region (Northeast Brazil) and 0.55 $kgTS/m^2{\cdot}d$ in a milder climate (Southeast Brazil) (Aisse *et al.*, 1999).

The solids loading to be applied is a function of the desired operating conditions of the drying bed, that is, sludge cake final moisture, sludge drying cycle and height of the sludge layer. The Brazilian standards previously mentioned (NB-570, ABNT, 1989) recommend a 15 $kgTS/m^2$ loading rate. Nevertheless, experimental results with higher rates using sludges from ponds were obtained and considered satisfactory.

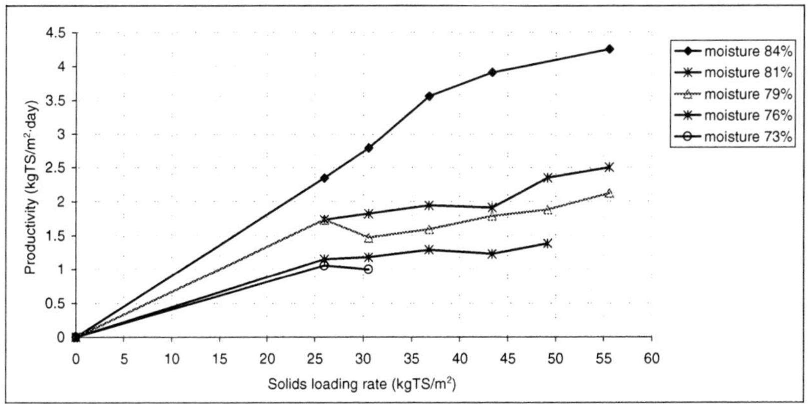

Figure 5.8. Productivity of sludge drying beds as a function of the applied solids loading (for a particular anaerobic pond sludge)

Example 5.5 (Continued)

Considering in this example a solids loading rate SLR_{sludge} of 30 kgTS/m^2 and a 73% cake final moisture, the productivity is derived from Figure 5.8 as:

$$\mathbf{P = 1.0\ kgTS/m^2{\cdot}d}$$

(c) Determination of the drying cycle

The time needed to promote dewatering (drying cycle) depends on the applied solids loading and the expected bed productivity.

$$T = SLR_{sludge}/P$$
$$T = 30\ (kgTS/m^2)/1.0\ (kgTS/m^2{\cdot}d) = \mathbf{30\ days}$$

(d) Required drying bed area

$$A = M_{sludge}/(P \times T)$$

where:

A = total area of the drying beds (m^2)
M_{sludge} = sludge mass (kgTS)
P = drying bed productivity (kgTS/m^2·d)
T = drying cycle (days)

$$A = 47 \times 10^3 (kgTS)/(1.0 kgTS/m^2{\cdot}d \times 30\ d) = \mathbf{1{,}567\ m^2}$$

The per capita required area is:

Per capita area = 1,567 m^2/20,000 inhabitants = 0.078 m^2/inhabitant

For the conditions of the present example, the sludge height to be applied on the drying beds is:

$$H_{sludge} = V_{sludge}/A$$
$$H_{sludge} = 579\ m^3/1{,}567\ m^2 = \mathbf{0.37\ m}$$

The drying bed area could be split according to the following alternatives (among others):

- 3 beds with 15 m × 30 m
- 5 beds with 12 m × 30 m
- 6 beds with 10 m × 30 m
- 7 beds with 10 m × 25 m

(e) Stagewise removal of the sludge from the pond

Since this sludge is removed from a stabilisation pond, it may remain on the drying beds for long periods, reaching even lower water contents, considering the usual large time intervals between successive sludge removal operations.

Example 5.5 (*Continued*)

In case the selected sludge removal technique allows withdrawals in two different stages, a feasible alternative to reduce the area requirements would be to remove and dry part of the sludge and, after the drying cycle, remove and dry the remainder of the sludge. In the above example, if half of the sludge is removed from the pond, the area would also be reduced by half.

(f) Influence of rainfall

The effect of rainfall on the removal of the sludge water was not taken into account when analysing the productivity presented in Figure 5.8. To consider rainfall, an estimate is needed about the average water removal rate from the drying bed. If P is the productivity of the bed and w_i and w_f represent the initial and final sludge moisture, respectively, the average water removal rate could be defined as:

$$Tw = Tw_i - Tw_f$$

where:

Tw_i = water loading rate = $P \times w_i/(1 - w_i)$
Tw_f = water withdrawal rate = $P \times w_f/(1 - w_f)$

Hence:

$Tw = P\{[w_i/(1 - w_i)] - [w_f/(1 - w_f)]\}$
$Tw = 1.0\{[0.92/(1 - 0.92)] - [0.73/(1 - 0.73)]\}$
$Tw = 8.80\ l/m^2 \cdot day = 8.80\ mm/day = 3{,}212\ mm/year$

Assuming 1,254 mm/year of rainfall in this period, the ratio of the water removal rate over the accumulated precipitation during the year is approximately 2.6 (= 3,212/1,254). The area of the drying bed should therefore be increased by 1/2.6 (38%) if rainfall is to be considered. The corresponding productivity shall also be reduced in 38%. Thus, the final area of the drying bed will become $A = 1{,}567\ m^2 \times 1.38 = 2{,}163\ m^2$.

5.5.4 Operational aspects

The solutions for operational problems in sludge drying beds are simple, as a consequence of their inherent conceptual simplicity. Table 5.10 presents some operational measures for the solution of drying bed problems.

5.6 CENTRIFUGES

5.6.1 General description of the process

Centrifugation is a process of forced solid/liquid separation by centrifugal force. In the first stage known as clarification, sludge solids particles settle at a much

Table 5.10. Main problems and solutions in the operation of sludge drying bed

Detected problem	Possible cause of the problem	Check, monitor, control	Solution
Lengthy dewatering cycle	Excessive sludge height applied to the drying beds	Recommended sludge height should be lower than 20–30 cm for satisfactory results	Remove dried sludge and thoroughly clean the drying bed. Apply a thin layer of sludge and measure its height reduction after 3 days. Apply the double of the verified height reduction in the third day after the first application
	Sludge application after improper cleaning of the drying bed	Check the cleaning condition (maintenance) of the drying beds	Remove the sludge after drying. Thoroughly clean the bed surface and replace the top sand layer with 12–25 mm of clean sand, if necessary
	Clogged drainage system, or broken piping		Make a slow countercurrent cleaning through the drying bed, connecting a clean water source into the bottom draining pipe. Check and replace filter media, if necessary. Completely drain the top layer to keep freezing from happening in cold seasons
	Undersized bed area	Try improve filtering with polymers	Typical polymer dosage is 2.3–13.6 g cationic polymer/kg dry solids. Significant dewatering rates improvement may result
	Climatic conditions of the region	Temperature, rainfall	Protect bed against adverse weather
Sludge feeding pipe clogged	Deposits of solids or sand in piping		Fully open the valves in the beginning of the sludge application for pipe cleaning. Apply water jets, if necessary
Very thin sludge being withdrawn from the digester tank	Separation problems in the digester, with excessive supernatant removal		Reduce sludge withdrawal rate from the digester
Flies on top of the sludge layer			Crack the sludge top-crust layer and pour a calcium borate larvicide or similar. Exterminate adult flies with insecticide
Foul odour when sludge is applied	Inadequate sludge digestion	Operation and digestion process	Adjust the operation of the digestion process
Lumps and dust with dewatered sludge	Excessive dewatering	Check water content	Remove sludge from bed when 40% to 60% water content is obtained

Source: WEF (1996)

higher speed than they would simply by gravity. In the second stage, compaction occurs, and the sludge loses part of its capillary water under the prolonged action of centrifugation. The cake is removed from the process after this latter stage.

As in sludge thickening, centrifugation is a sedimentation process originating from the difference of density between a particle and the surrounding liquid. The process may be described by Stokes equation, which expresses the settling velocity of a solid particle in a fluid:

$$V = [g \cdot (\rho_S - \rho_L) \cdot d^2]/(1{,}800 \cdot \mu) \qquad (5.2)$$

where:

V = settling velocity of the solid particle in the liquid (m/s)
g = gravitational constant (m/s^2)
ρ_S = particle density (kg/m^3)
ρ_L = liquid density (kg/m^3)
d = average particle diameter (m)
μ = liquid viscosity (kg/m·s)

The equation above shows that the settling velocity of a particle is directly proportional to the difference between the particle and the liquid densities and to the square of the particle diameter, and inversely related to the liquid viscosity.

The acceleration resulting from centrifugation (G) is usually related to the gravitational constant, as a multiple of g ($g = 9.81\ m/s^2$). The centrifugal acceleration over a particle in a liquid inside a cylinder is given by:

$$G = \omega^2 \cdot R = (2\pi N/60)^2 \cdot R \qquad (5.3)$$

where:

G = centrifugal acceleration of the particle (m/s^2)
ω = angular velocity (rad/s)
R = radius (m)
N = rotation speed (revolutions/min)

The settling velocity V of a solid particle in a centrifuge with radius R is obtained replacing *g* in Equation 5.2 by *G* defined in Equation 5.3.

A sludge dewatering centrifuge works with a centrifugal acceleration 500–3,000 higher than the gravitational constant, and the settling velocity of 10 m/hour is 50 times greater than the natural thickening velocities for sludges. The magnitude of the forces involved makes the inner bond forces among particles to split. This allows a better separation than would be possible through a simple static settling.

5.6.2 Types of centrifuges

Centrifuges may be used indistinctly for sludge thickening and dewatering. The operating principle remains the same, with the possibility of installing centrifuges in series, with thickening being accomplished in the first stage and dewatering in the following one. Equipment capacities vary, but a range from 2.5 m^3/hour to 180 m^3/hour of incoming sludge flow is usually available.

Vertical and horizontal shaft centrifuges are used in sludge dewatering. They differ mainly in the type of sludge feeding, intensity of the centrifugal force applied and manner of discharging the cake and the liquid from the equipment. Horizontal shaft centrifuges are most widely applied for thickening and dewatering of sludges. A relatively lower cake solids contents and the need to feed semi-continuously are among the reasons why vertical shaft centrifuges are less used. The advantages and disadvantages of horizontal-shaft centrifuges are presented.

Horizontal centrifuges in use today are of solid-bowl type, with moving parts consisting of the rotating bowl and the rotating conveyor/scroll, made of stainless steel or carbon steel. The main components of a centrifuge are: support basis, bowl, conveyor scroll, cover, differential speed gear, main drive and feeding pipe.

The support basis, normally built of steel or cast iron, has vibration insulators for reducing vibration transmission. The cover, which involves all moving parts, helps to reduce odours and noise and collects both the centrifuged liquid (centrate) and the dewatered sludge. The bowl has a cone-cylinder shape, with variable characteristics depending upon the manufacturer. The bowl length:diameter ratio varies from 2.5:1 to 4:1, with diameters ranging from 230 to 1,800 mm. The differential speed gear allows the rotational speed differences between the bowl and the screw conveyor.

Both the bowl and the conveyor scroll rotate in the same direction at high speed, with the scroll speed slightly different from the bowl speed, allowing for a conveying effect to take place. The scroll is located inside the bowl core, keeping a 1–2 mm radial aperture, just enough for the passage of the centrifuged liquid. Velocities between 800–3200 rpm normally yield cakes containing solid levels greater than 20% and a clarified centrate. Higher centrifugation speeds imply lower polyelectrolyte consumption, higher solids capture and possibly higher cake solid content. This may come at the expense of higher maintenance costs due to bearings abrasion.

The centrifuge reduction gear-box produces rotational differences from 1 to 30 rpm between the bowl and the screw conveyor. Sludge feeding rate, the speed difference and the bowl rotational speed are the main parameters controlling the solids retention time inside a centrifuge. High solids cakes result from high retention time, low speed differential and compatible sludge feeding rate.

The sludge thickening achieved depends upon the sludge type and the initial solids concentration. In general, it can be said that dryer cakes are obtained when the speed difference between the bowl and the screw conveyor is kept to a minimum.

Horizontal centrifuges may be classified according to the direction of the feeding flow and the way the cake is withdrawn into *co-current* or *counter-current*. They

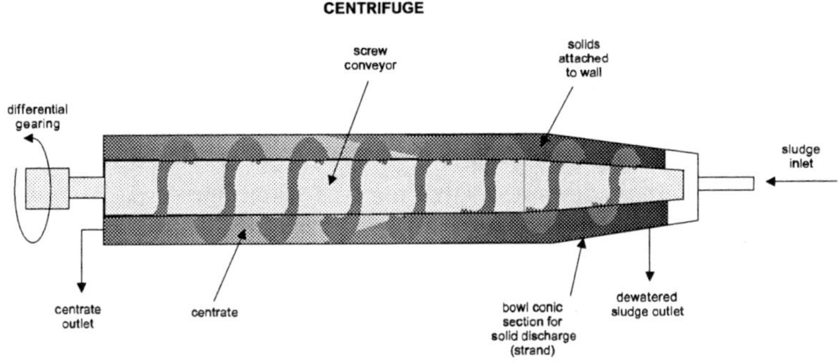

Figure 5.9. Diagram of a countercurrent horizontal centrifuge

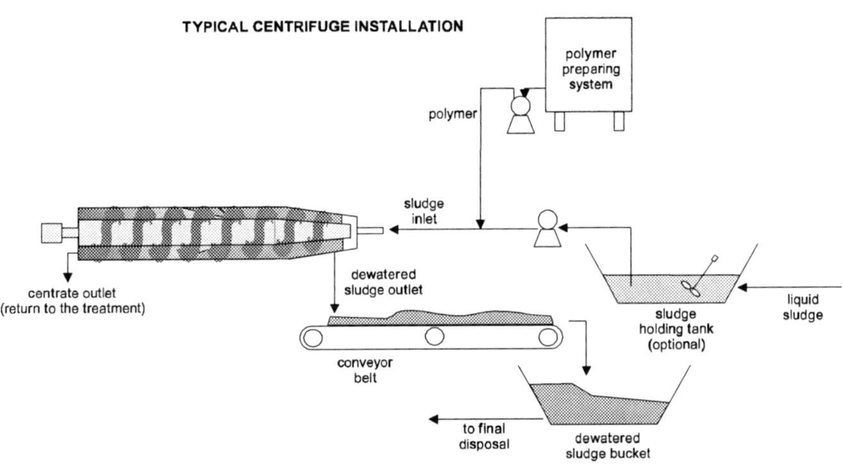

Figure 5.10. Typical installation of a decanter-type centrifuge

differ in the sludge feeding points, the way the centrate is removed (liquid phase) and the direction of the solid and liquid phases within the unit. In co-current centrifuges, the solid phase and the liquid phase cross all the way through the longitudinal axis of the centrifuge bowl until being discharged. In counter-current models, the sludge is fed on the opposite side from the centrate releasing point. The solid phase is routed out by the screw conveyor towards the end of the conic section, while the liquid phase makes the opposite path. Figure 5.9 shows a diagram of a counter-current centrifuge.

5.6.3 Dewatering flowsheet using centrifuge

Figure 5.10 illustrates the flowsheet of sludge dewatering by centrifuge. The area required for a large centrifuge having a sludge feeding capacity from 10 to 40 L/s is approximately 40 m^2 (WEF, 1996). This is one of the main advantages centrifuges have over other mechanical sludge dewatering equipments. Centrifuges do not emit aerosol or excessive noise, and may be installed in open sheds. Electric power consumption and maintenance costs are fairly high, limiting their use for wastewater treatment plants with flows higher than 100 L/s, or where area availability is a limiting factor. Other components that must be taken into consideration in the design of the dewatering facility are:

- conditioning system, with polymer tanks, dosage equipments and piping
- sludge dosing pumps and piping
- access of vehicles for centrifuge maintenance
- areas for circulation, ventilation, electric equipments and smell control

Depending upon the centrifuge operating regime, liquid sludge may be kept in a sludge holding tank, equalising sludge flow prior to dewatering. A similar comment is equally valid regarding other mechanised dewatering processes.

In a fully mechanised system, dewatered sludge is conveyed through a conveyor belt towards a container or the storage area within the treatment plant where it will remain until transportation to final disposal. The conveyor belt mechanism should be set to switch on just before the centrifuge, and to switch off a few minutes after the centrifuge.

5.6.4 Performance

The characteristics influencing centrifugation performance are the same that influence sedimentation. The main variables influencing centrifuge performance are sludge solids concentration, type of conditioning, feed flow and temperature. Larger particles are easily captured by the centrifuge, while finer particles require conditioning to reach a sufficient size for capture.

The effectiveness of chemicals such as polyelectrolytes is more closely related with the solids concentration of the centrate (solids capture) than with the cake solids content. Cationic polyelectrolyte is often used as a flocculation aid, giving better solids capture and greater feed flow. Anionic polyelectrolytes are used along with metallic coagulants.

Another factor determining the centrifuge efficiency is the sludge volatile solids concentration. High sludge stabilisation levels improve centrifuge performance, allowing high cake solids content.

Several mechanical factors influence the equipment performance. The manufacturer, however, is responsible for most of the settings. Plant staff shall undertake the following adjustments:

- pool depth, normally set by the supplier after preliminary tests
- injection point of the metallic coagulants and polyelectrolyte
- feed flow
- bowl and conveyor scroll differential speed

Table 5.11. Typical centrifuge performance in sludge dewatering

Type of sludge	TS concentration in cake (%)	Solids capture (%)	Polyelectrolyte dosage (g/kg)
Primary raw sludge	28–34	95	2–3
Anaerobic sludge	35–40	95	2–3
Activated sludge	14–18	95	6–10
Raw mixed* sludge	28–32	95	6–10
Anaerobic mixed sludge	26–30	95	4–6
Aerobic sludge**	18–22	95	6–10

* primary sludge + excess activated sludge
** extended aeration or excess activated sludge

Table 5.12. Suggestions of capacities and number of centrifuges

Liquid sludge flow (m^3/d)	Operating hours	Number of units		Capacity of each unit (m^3/hour)
		In operation	Spare	
40	7	1	1	6
80	7	1	1	12
350	15	2	1	12
800	22	2	1	18
1,600	22	3	2	25
4,000	22	4	2	45

- Number of operating hours = Sludge flow (m^3/d)/[(Number of operating units) × (Capacity of each unit, in m^3/hour)]
- Refer to manufacturer's catalogue for different centrifuge capacities

Source: Adapted from EPA (1987)

Adjustment of a centrifuge may be done either aiming at a drier cake production or a better-quality centrate, as the operator requires. Emphasis on cake solids content increases centrate solids concentration (low capture of solids), and vice-versa. Table 5.11 shows typical performance data of horizontal axis centrifuges in sludge dewatering.

5.6.5 Design

The sizing of a centrifuge dewatering facility is based upon manufacturer data on the equipment loading capacity and the type of liquid sludge. Whenever possible, preliminary tests should be carried out. Manufacturers should always be consulted on design details and characteristics of the different models. Dimensions of the equipments (diameter and length) vary among suppliers, as well as performance data such as power, maximum bowl speed and maximum centrifugal force.

The number of operating and spare units is a function of plant capacity and sludge production, as well as maintenance staff size and availability of alternative sludge disposal routes. General guidelines to select the number of operating and spare units presented in Table 5.12, are based on EPA (1987). These values, however, may vary widely from one case to another.

The main information required for sizing is:

- type of sludge to be treated
- daily sludge flow
- dry solids concentration

Other useful data used for predicting the performance of a centrifuge are the SVI (Sludge Volume Index) and fixed and volatile solids content.

Example 5.6

For the wastewater treatment plant of Example 2.4 (conventional activated sludge plant, with anaerobic digestion of mixed sludge; population = 67,000 inhabitants), size the centrifuges for sludge dewatering. Effluent sludge from secondary digester (influent to the dewatering unit) is 46.9 m^3/day.

Solution:

(a) Influent sludge flow

$$Q_{av} = 46.9\ m^3/d = 1.95\ m^3/hour = 1{,}950\ L/hour$$

Considering an hourly peak factor of 1.5, the maximum sludge flow to be dewatered is:

$$Q_{max} = 1.5 \times 1.95\ m^3/hour = 2.93\ m^3/hour = 2{,}930\ L/hour$$

(b) Equipment selection

The selection of the centrifuge can be done considering the maximum sludge flow to be dewatered. This information, together with other relevant data, shall be supplied to the manufacturer for the final selection.

For this particular example, and based on Table 5.12, a 6 m^3/hour centrifuge is selected (an operating unit and a spare one).

(c) Operating hours

The following number of operating hours per day is computed for the average influent sludge flow:

$$\text{Operating time (hour/d)} = \frac{\text{Average influent sludge flow } (m^3/d)}{\text{Number of units} \times \text{Unitary capacity } (m^3/hour)}$$

$$= \frac{46.9\ m^3/d}{1 \times 6\ m^3/hour} = 8\ hours/d$$

In case the production of sludge to be dewatered is continuous (24 hours/d), a sludge holding tank capable of storing the sludge during non-operating hours is necessary. The liquid sludge in the current example is being withdrawn from the secondary digesters, which already play the role of a sludge holding tank.

Table 5.13. Problems and solutions in centrifuge operations

Operational problem	Consequence	Solution
Inadequate material blades	Excessive abrasion	Replace with more resistant material
Rigid feeding pipes	Pipe cracks and nipple leaks	Replace with flexible pipes
Grit in the sludge	Excessive abrasion of the equipment	Either review operation or install grit chamber
Excessive vibrations	Destabilisation of electric and mechanical parts	Install adequate shock absorbers
Electric control panels in the same room	Corrosion and deterioration of controls	Move electric panels to different room

5.6.6 Operational aspects

Variables affecting centrifuge performance may be classified into three categories, similarly as with other mechanical dewatering equipment:

- sludge characteristics
- sludge conditioning (preparation)
- equipment mechanical setting

A troubleshooting guide is presented in Table 5.13 regarding centrifuge operation.

5.7 FILTER PRESS

5.7.1 General description of the process

Filter presses were developed aiming at industrial use and later underwent changes to make them suitable for wastewater sludge dewatering operations. They operate through batch feeding which demands skilled operators. The major quality of filter presses is their reliability. The main advantages of filter press are:

- cake with higher solids concentration than any other mechanical equipment
- high solids capture
- quality of the liquid effluent (filtrate)
- low chemical consumption for sludge conditioning

5.7.2 Working principle

The filter press operating cycle varies from 3–5 hours, and may be divided into three basic stages:

- *Filling*. Pumped sludge is admitted into empty gaps between consecutive filter plates. The filling period may reach 20 minutes, but usual time intervals are 5 to 10 minutes. The filling pump pressure is sufficient to immediately initiate the solid/liquid separation processes in filter cloths.

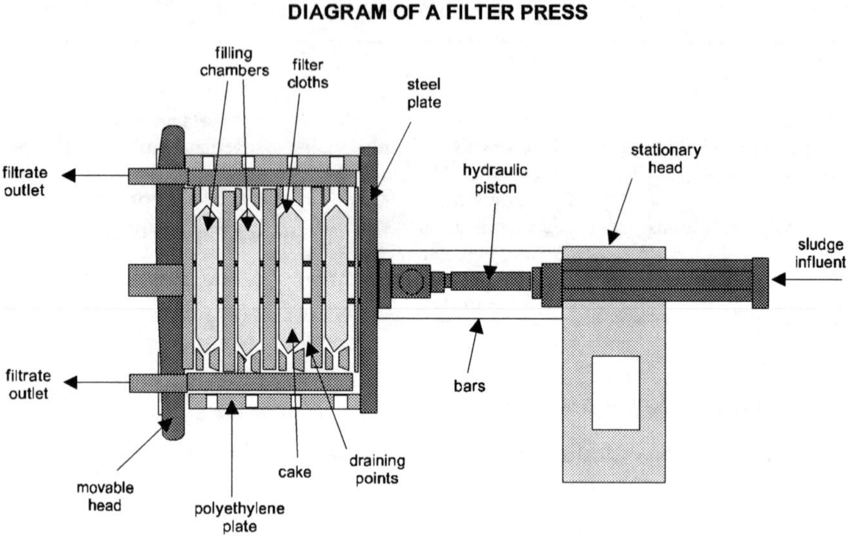

Figure 5.11. Operating diagram of a filter press

- *Filtration under maximum pressure.* During the filtration phase, the applied pressure may reach 170 kPa (17 atm –250 psi).
- *Cake discharge.*

The time for each batch varies according to sludge feeding pump flow, type of sludge, sludge solids content, influent sludge filterability and cleaning status of the filter cloth.

Figure 5.11 shows diagrammatically a cross section of a filter press. The liquid sludge is pumped into the recessed plates, enveloped by filter cloths. Pumping of the sludge increases the pressure in the space between plates, and the solids (filter cake) are left attached to the media as the liquid sludge passes through the filter cloth.

Afterwards, a hydraulic piston pushes the steel plate against the polyethylene plates, compressing the cake. Both the movable and the stationary head have support bars specifically designed for this purpose.

The filtrate passes through the filter cloths and is collected by draining points and filtrate channels. The filtrate usually has less than 15 mg SS/L. The cake is easily removed from the filter as the pneumatic piston moves back and the plates are separated from each other. After falling down from the plate, the compressed cake is ready to be routed to storage or final destination.

Figure 5.12 presents a typical filter press installation for sludge dewatering.

Nowadays, filter presses are automated, greatly reducing the need for hand labour. The weight of the equipment, its initial costs and the need for regular filter cloth replacement make the use of filter presses restricted to medium size and large wastewater treatment plants.

Table 5.14. Optimal performance of filter presses in sludge dewatering

Type of sludge	Solids content in cake (%)	Cycle (hours)
Primary	45	2.0
Primary + activated sludge	45	2.5
Activated sludge	45	2.5
Anaerobic primary	36	2.0
Anaerobic + activated	45	2.0

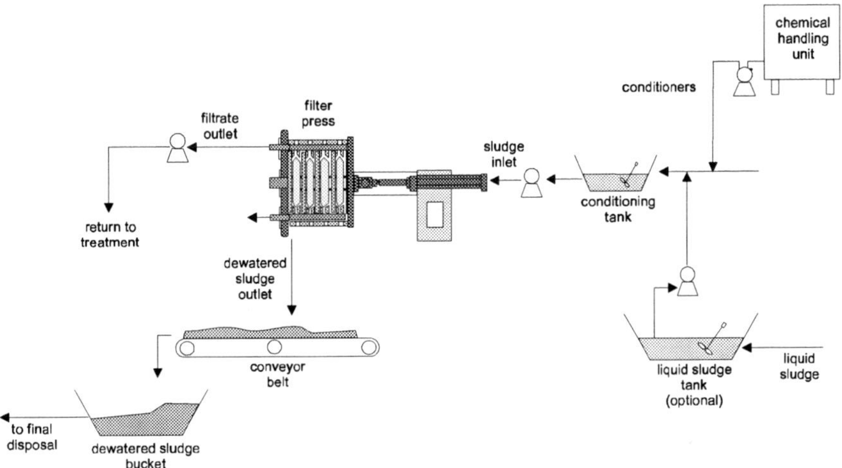

Figure 5.12. Flowsheet of a filter press facility for sludge dewatering

5.7.3 Performance

Table 5.14 presents optimal performance values for filter presses.

5.7.4 Design

The sequence of calculations for filter press sizing is illustrated in Example 5.7.

Example 5.7

For the wastewater treatment plant of Example 2.4 (conventional activated sludge, with anaerobic digestion of mixed sludge; population = 67,000 inhabitants), size a filter press system for sludge dewatering. The effluent sludge from the secondary digester (influent to dewatering) has a solids load of 1,932 kgTS/d (81 kg/hour) and a solids content of 4.0% TS (specific gravity = 1.03). The

Example 5.7 (_Continued_)

dewatering period shall be 5 days/week and 8 hours/day. TS content in cake must reach 40% TS. Estimated cake specific gravity is 1.16. Specific operating conditions are as follows:

- Cake thickness = 30 mm
- Filter press operating pressure = 15 bars
- Chemical conditioner: lime and ferric chloride
- Chemical dosing:
  - 10 to 20% CaO (average of 15%)
  - 7.5 % $FeCl_3$
- Chemical sludge formed:
  - Lime = 80% dosed CaO
  - $FeCl_3$ = 50% dosed $FeCl_3$
- Peak coefficient in sludge production = 1.25

Solution:

(a) Sludge production rate

- Sludge mass (dry basis) = 1,932 (kgTS/d) × 7 (d/week) = 13,524 kgTS/week = 13.5 tonne TS/week
- Wet sludge volume = 13.5 (tonne/week) / (0.04 TS × 1.03) = 328 m^3/week

(b) Daily and hourly demand for dry solids processing without chemicals

Based on a 5-day/week and 8-hour/day operating schedule, one has:

- Daily rate = 13.5 (tonne/week) ÷ 5 (d/week) = 2.7 tonne/day
- Hourly rate = 2.7 (tonne/d) ÷ 8 (hours/d) = 0.34 tonne/hour or 340 kg/hour

(c) Daily cake volume considering chemicals added

(c1) Solids mass in cake (kg/d):

$$\text{average} = M_S = 1{,}932\ \text{kg/d}$$
$$\text{maximum} = M_{S,max} = 1.25 \times 1{,}932 = 2{,}415\ \text{kg/d}$$

(c2) Average dosage of $FeCl_3$ (kg/d)

$$M_{FeCl_3} = 10^{-2} \times (\%_{FeCl_3}) \times M_S$$
$$\text{average} = M_{FeCl_3} = 10^{-2} \times 7.5 \times 1{,}932 = 145\ \text{kg/d}$$
$$\text{maximum} = M_{FeCl_3,max} = 10^{-2} \times 7.5 \times 2{,}415 = 181\ \text{kg/d}$$

(c3) Chemical sludge mass $FeCl_3$ (kg/d)

$$M_{S,FeCl_3} = (L_{FeCl_3} \cdot M_{FeCl_3})$$
$$\text{average} = M_{S,FeCl_3} = 0.5 \times 145 = 73\ \text{kg/d}$$
$$\text{maximum} = M_{S,FeCl_3,max} = 0.5 \times 181 = 91\text{kg/d}$$

Example 5.7 (Continued)

(c4) Average lime dosage

$$M_{CaO} = 10^{-2} \times (\%_{CaO}) \times M_S$$
$$\text{average} = M_{CaO} = 10^{-2} \times 15 \times 1{,}932 = 290\ kg/d$$
$$\text{maximum} = M_{CaO,max} = 10^{-2} \times 15 \times 2{,}415 = 362\ kg/d$$

(c5) Chemical sludge mass CaO (kg/d)

$$M_{S,CaO} = (L_{CaO} \cdot M_{CaO})$$
$$\text{average} = M_{S,CaO} = 0.8 \times 290 = 232\ kg/d$$
$$\text{maximum} = M_{S,CaO,max} = 0.8 \times 362 = 290\ kg/d$$

(c6) Total solids mass (kg/d)

$$M_s\ \text{total} = M_S + M_{FeCl_3} + M_{S,FeCl_3} + M_{CaO} + M_{S,CaO}$$
$$\text{average} = M_s = 1{,}932 + 145 + 73 + 290 + 232 = 2{,}671\ kg/d$$
$$\text{maximum} = M_{s,max} = 2{,}415 + 181 + 91 + 362 + 290 = 3{,}339\ kg/d$$

(c7) Cake solids concentration

$$\text{average} = C_{ST} = 40\%$$
$$\text{maximum} = C_{ST,max} = 40\%$$

(c8) Cake specific gravity

$$\text{specific gravity} = 1.16$$

(c9) Cake volume (m^3/d)

$$\text{average} = V_s = 2{,}671\ (kgTS/d)/(40(\%) \times 1.16) = 5{,}757\ L/d = 5.8\ m^3/d$$
$$\text{maximum} = V_{s,max} = 3{,}339\ (kgTS/d)/(40(\%) \times 1.16)$$
$$= 7{,}196\ L/d = 7.2\ m^3/d$$

(c10) Daily and hourly demand of dry solids processing with chemicals

Based on a 5-day/week and 8 hour/day operational schedule, one has:

- Average daily rate = 2,671 (kgTS/d) × 7 (d/week) ÷ 5 (d/week) = 3,739 kgTS/d = 3.8 tonne/d/
  = 3,739 (kgTS/d)/(40(%) × 1.16) = 8,060 L/d = 8.1 m^3/d
- Maximum daily rate = 3,339 (kgTS/d) × 7 (d/week) ÷ 5 (d/week) = 4,675 kgTS/d = 4.7 tonne/d/
  = 4,675 (kgTS/d)/(40(%) × 1.16) = 10,075 L/d = 10.1 m^3/d

Example 5.7 (Continued)

(d) Daily production of cakes

(d1) Cake volume per filter-press plate

Adopting 1.0 m × 1.0 m plate size and 30 mm cake thickness, the cake volume per plate is:

$$V_p = 1.0 \times 1.0 \times 0.03 = 0.030 \text{ m}^3/\text{plate}$$

Assume for safety: $V_p = 0.025$ m^3/ plate

(d2) Daily cake production

The daily number of cakes is calculated by dividing the total daily cake volume by the cake volume of one plate (V_p):

$$\text{Average:} N_c = (V_t/V_p) = 8.1/0.025 = 324 \text{ cakes/day}$$
$$\text{Maximum:} N_{c\,max} = 10.1/0.025 = 404 \text{ cakes/day}$$

(e) Required number of filter plates

The required number of filter plate units and the number of filtering cycles needed shall cope with 324 cakes under normal operation and 404 cakes during sludge peak production. Cake production as a function of the number of filter press units, number of plates and filtration cycles may be selected from the chart that follows:

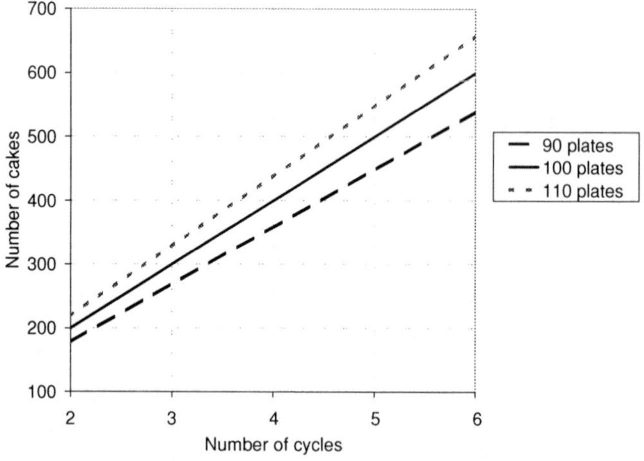

Amount of cakes produced as a function of the number of filtration cycles per day

From the chart, the following combination is able to cope with the average demand:

- One filter press with 110 plates (should a lower number of plates be selected, the number of cycles would increase with a shorter operational time length).

Example 5.7 (Continued)

(f) Duration of pressing cycles

The number of cycles to cope with demand is:

$$N\ cycles = \frac{N_c}{N_f \cdot N_p}$$

where:

N cycles = number of cycles
N_c = daily cake production (number of cakes/day)
N_f = number of filter presses
N_p = number of plates

The cycles per filter should have the following characteristics:

average: N cycles = 324/(1 × 110) = 2.95 cycles/filter·d
= 3 cycles/filter·d
time of cycle = 8/3 = 2.7 hours/cycle
maximum: N cycles = 404/(1 × 110) = 3.67 cycles/filter·d
= 4 cycles/filter·d
time of cycle = 8/4 = 2.0 hours/cycle

As a 2-hour cycle time is very short, at least 5 daily cycles shall be necessary to meet the maximum forecasted demand requirements.

5.7.5 Operational aspects

Filter press performance varies with the fed sludge properties and operational adjustments in the equipment control parameters. The following parameters may be adjusted by the operator:

- working pressure: according to supplier and type of equipment, the working pressure may vary within either one of the following ranges: 656 to 897 kPa (6.5–8.9 atm) or 1,380 to 1,730 kPa (13.6–17.1 atm)
- sludge feeding rate
- total filtration periods: including interim periods to operate in different pressure levels, when pressure variations are allowed during the operating cycle
- type of filter cloth: has direct influence on equipment performance
- type of filter plates: steel plates are thinner and stronger, producing a larger dewatered sludge mass per filtration cycle. Polypropylene plates are cheaper and resistant to corrosion; however, as they are 50% thicker than steel plates, their use causes a reduction in the number of plates per filter press, with a consequent reduction in the cake production per filtration cycle.

Filter press operation requires careful visual inspection before any filtration cycle begins. The operator should ensure that all filter cloths are duly coupled without any folds and free from dirt. Torn filter cloths must be replaced. Tearing usually occurs around its central portion or at intermediate anchor points. No object shall be placed between or upon plates when the operation cycle is taking place. At the end of any working shift, the equipment must be washed up and conditions of filter cloths checked.

5.8 BELT PRESSES

5.8.1 General description of the process

Belt presses, also named **belt-filter presses**, may be divided into three distinct zones (a) gravity-dewatering zone, (b) low-pressure zone, (c) high-pressure zone.

The gravity-dewatering zone is located at the equipment entrance, where sludge is applied on the upper belt and the free water percolates through the cloth pores. Next, the sludge is routed to a low-pressure zone (also known as wedge zone), where it is gently compressed between the upper and lower belts, releasing the remainder of the free water. Within the high-pressure zone, formed by several rollers of different diameters in series, the sludge is progressively compressed between two belts, releasing interstitial water. Dewatered sludge is then removed by scrapers located on the upper and lower belts. The upper and lower belts are washed by high-pressure water jets before receiving fresh diluted sludges. The cloth washing water must have a minimum pressure of 6 kg/cm^2 and sufficient flow to remove attached sludge and polyelectrolyte residues from the cloth.

As belt presses are open, they have the disadvantage of aerosol emission, high noise level and possible foul odour emission (depending on the type of sludge). The high number of bearings (40–50 depending on the manufacturer) is another significant disadvantage of belt presses, as they require regular attendance and replacement. As advantages though, they have low initial costs and reduced electric power consumption. Recent developments in decanter-type centrifuges triggered intense competition among suppliers of both types of dewatering equipment. Despite their higher initial costs, centrifuges are being favoured so far.

Figure 5.13 presents the schematics of a typical belt press installation.

5.8.2 Performance

Typical performance of belt presses for different types of sludge can be seen in Table 5.15. A comparison with Table 5.14 (filter presses) shows that belt presses produce a cake with higher water content, for the same type of sludge.

5.8.3 Design

Belt widths are commercially available in the 0.5 m–3.5 m range (most common size is 2.0 m). Usual hydraulic loads in terms of belt width range from 1.6 to 6.3 L/s·m, varying with sludge characteristics and desired dewatering efficiency. The solids loading rates range from 90 to 680 kgTS/m·hour.

Table 5.15. Typical performance of belt presses

Type of sludge	Hydraulic load (m^3/hour)	Solids load (kg/hour)	Solids concentration in liquid sludge (% TS)	Solids concentration in the cake (% TS)	Solids capture (%)
Anaerobic*	6.4–15	318–454	3–5	18–24	95
Aerobic**	7.3–23	181–318	1–3.0	14–18	92–95
Activated sludge	10.4–23	136–272	0.5–1.3	14–18	90–95
Raw primary	11.4–23	681–1,134	4–6	23–25	95
Raw mixed	9.1–23	454–681	3–5	23–28	95

* 50% primary/50% activated sludge in weight
** aerobically digested activated sludge

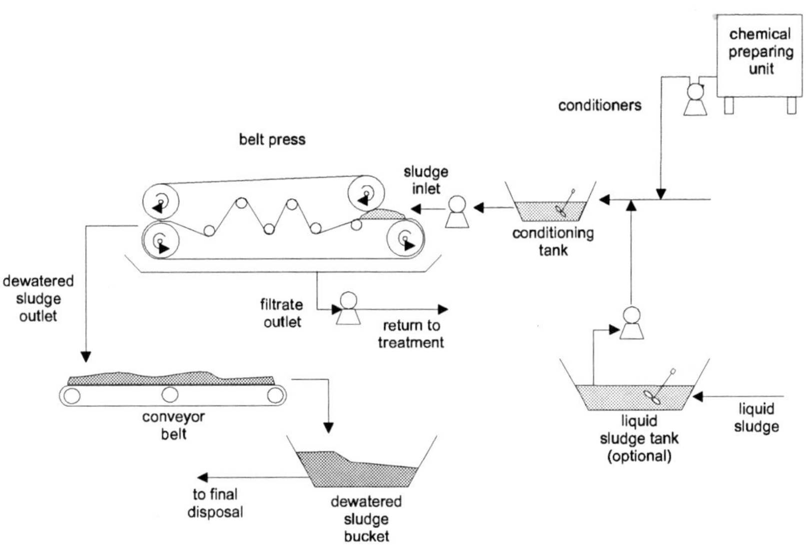

Figure 5.13. Flowsheet of a belt press installation

Example 5.8

For the wastewater treatment plant of Example 2.4 (conventional activated sludge, with anaerobic digestion of mixed sludge; population = 67,000 inhabitants), size the sludge belt press. Effluent sludge from secondary digester (influent to dewatering) has a solids load of 1,932 kgTS/d (81 kg/hour) and a solids content of 4.0% TS (specific gravity = 1.03). The dewatering period shall be 5 days/week and 8 hours/day. Other data to be considered are as follows:

- solids content in the cake = 25%
- nominal belt capacity = 272 TS/hour·m
- filtrate TS concentration = 900 mg/L

Example 5.8 (***Continued***)

- washing water flow = 1.51 L/s·m of belt
- cake specific gravity = 1.07
- filtrate specific gravity = 1.01

Solution:

(a) Sludge production

- Sludge load (dry basis) = 1,932 (kgTS/d) × 7 (d/week) = 13,524 kgTS/week = 13.5 tonne TS/week
- Wet sludge = 13.5 (tonne/week) ÷ (0.04 TS × 1.03) = 328 m^3/week

(b) Daily and hourly demand for dry solids processing

Based on a 5-day/week and 8-hour/day operating schedule, one has:

- Daily rate = 13.5 tonne/week) ÷ 5 (d/week) = 2.7 tonne/d = 2,700 kg/d
- Hourly rate = 2.7 (tonne/d) ÷ 8 (hours/d) = 0.34 tonne/hour or 340 kg/hour

(c) Belt press size

Belt width:

$$B = (\text{hourly rate}) \div (\text{nominal load}) = 340\ (\text{kg/hour}) \div 272\ (\text{kg/hour·m}) = 1.25\ \text{m}$$

One 1.5 m width belt press plus a spare unit will be adopted.

(d) Filtrate flow based on solids and flow balances

(d-1) Solids balance

$$\begin{aligned} \text{Solids in the sludge} &= \text{Solids in the cake} + \text{Solids in filtrate} \\ 2{,}700\ (\text{kg/d}) &= (S \times 1.07 \times 0.25) + (F \times 1.01 \times 0.0009) \\ &= 0.268\,S + 0.0009\,F = 2{,}700 \end{aligned}$$

where:
F = filtrate flow (L/day)
S = cake flow (L/day)

(d-2) Water balance

$$\text{Sludge flow} + \text{Washing flow} = \text{Filtrate flow} + \text{Cake flow}$$

$$\text{Daily sludge flow} = 328\ (m^3/\text{week}) \times (1/5)\ (\text{week/d}) = 65.6\ m^3/d$$

$$\text{Washing water flow} = 1.51\ (\text{L/s·m}) \times 1.5\ (\text{m}) \times 3{,}600\ (\text{s/hour}) \times 8\ (\text{hours/d}) = 65.2\ m^3/d$$

Example 5.8 (Continued)

$$65.6 + 65.2 = F + S$$
$$F + S = 130.8\,(m^3/d) = 130{,}800\,(L/d)$$

(d-3) Solution of the equation system

Combining Equations d-1 and d-2 above, one has:

$$0.268\,S + 0.0009\,F = 2{,}700$$
$$F + S = 130{,}800$$
$$F = 130{,}800 - S$$

Replacing it in the second equation:

$$0.268\,S + 0.0009\,(130{,}800 - S) = 2{,}700$$
$$0.268\,S + 117.7 - 0.0009\,S = 2{,}700$$
$$0.2671\,S = 2{,}582$$
$$S = 9{,}667\,(L/d)\ \text{(cake flow)}$$

Therefore:

$$F = 130{,}800 - 9{,}667 = 121{,}133\ L/d\ \text{(filtrate flow)}$$

(e) Solids capture

$$\text{Solids capture (\%)} = \frac{(TS_{sludge}) - (TS_{filtrate})}{TS_{sludge}} \times 100 =$$

$$\text{Capture (\%)} = 100 \times [2{,}700\,(kg/d) - (121{,}133\,(L/d) \times 1.01 \times 0.0009\,(kg/m^3))]/2{,}700\,(kg/d)$$
$$\text{Capture (\%)} = 96\%$$

(f) Operating conditions

Under normal circumstances, the operating time is:

$$\text{Operating time} = 1{,}932\,(kgTS/d)/[272\,(kg/hour.m) \times 1.5\,(m)] = 4.7\ \text{hours}$$

During peak daily production, working shift periods should be proportionately longer. For instance, if the peak daily factor is 1.5 (daily production equal to 1.5 times the average daily production), the operating time will be $1.5 \times 4.7 = 7.1$ hours. In case the spare unit is used, the operating time is reduced.

5.8.4 Operational aspects

The main variables influencing belt press performance are listed below and must be controlled by the plant operator:

- Solids content in the cake
- Solids loading rate
- Solids capture
- Hydraulic loading rate
- Belt speed
- Belt tension
- Type and dosage of polyelectrolyte
- Sludge solids concentration
- Flocculation velocity
- Point of application of polyelectrolyte
- Pressure and flow of belt washing water

Similar to other mechanical dewatering equipments, belt presses demand careful maintenance and need thorough cleaning at the end of every operating shift. Special care must be taken towards spray nozzles for belt cleaning and to the belt tracking and tensioning system. Spray nozzle cleaning frequency is directly dependent upon the quality of the service water being used. When recycled plant effluent is used, a filter must be installed to ensure that washing water is free of solids that could clog the spray nozzles.

As belt presses allow exposure of the sludge during the entire dewatering process, it is essential to assure adequate ventilation to reduce adverse environmental impacts and keep the risk of high hydrogen sulphide (H_2S) concentrations low when the facility is processing anaerobically digested sludge.

5.9 THERMAL DRYING

The thermal drying process is one of the most efficient and flexible ways of reducing cake moisture content from dewatered organic industrial and domestic sludges. Thermal drying may be used for different sludge types, either primary or digested, and a feeding sludge solids content of 15%–30% is recommended (obtained through prior mechanical dewatering). The removal of water can be controlled and final solids content shall be chosen depending upon the disposal route, for instance:

- *sludges addressed to incineration*: solids content in the range of 30–35% to ensure the autothermic operation
- *sludges addressed to landfill disposal*: solids content around 65%
- *biosolids addressed to farming through retail sale (unrestricted use)*: solids contents higher than 90%

Under ideal conditions, 2,744 kJ (655 kcal) of energy are needed to evaporate 1 kg of sludge water, and it is usual to increase this value up to 100% for normal operational conditions. The total energy demand will depend on the efficiency of the selected equipment and on the type of the processed sludge. Part of this energy must come from external sources, such as fuel oil, natural gas etc. Biogas generated in anaerobic digesters may constitute an ancillary energy source for thermal drying of wastewater sludges. The main advantages of sludge thermal drying are:

- significant reduction in sludge volume;
- reduction in freight and storage costs of the sludge;
- generation of a stabilised product suitable to be easily stocked, handled and transported;
- production of a virtually pathogen-free final product;
- preservation of biosolids fertilising properties;
- no requirements of a special equipment for land application;
- sludge is suitable for incineration or landfilling;
- product may be put into sacks and distributed by retail dealers.

Thermal drying has been historically adopted in retrofitted wastewater treatment plants that were already using some biological sludge stabilisation process, mainly anaerobic digestion. Its technology attracted considerable interest of designers and water companies especially in Europe aiming at the thermal drying of raw sludge. The suppression of the biological stabilisation stage significantly reduces capital costs, and favours the production of pellets with high organic matter content and heating value. These features add value to the product, furthering its use either in agriculture or as fuel source.

Thermal drying consists of sludge heating within a hermetically sealed environment, with evaporation and collection of the moisture. The sludge is taken out from the dryer as 2–5 mm average diameter pellets and solids content above 90% (when farming is being considered). The evaporated liquid is condensed and returns to the treatment plant headworks. The high temperature assures that the produced pellet is free of pathogens and qualified to be land-applied without restriction. The process is compact, completely enclosed, and does not allow release of foul odours. It is suitable for medium and large treatment plants with limited land availability and located next to residential areas.

Fuel consumption is the major operational component of thermal drying systems. Alternative fuel sources, such as natural gas or methane gas from anaerobic digesters or sanitary landfills, may lead to considerable reduction in operational costs. Selling the final product as class-A biosolid may reimburse a significant amount of the process expenses and help to balance the operational costs of the system.

It is important to point out that programmes for biosolids handling and resale in Europe or in North America have not yet been able to produce a positive financial balance. Sludge processing costs must be covered by water/wastewater rates.

Thermal drying is further discussed in Chapters 6 and 9.

6

Pathogen removal from sludge

Marcelo Teixeira Pinto

6.1 INTRODUCTION

As stated in Chapter 3, pathogenic organisms are sludge constituents that cause most concern in its processing and final disposal. Bacteria, viruses, protozoan cysts and intestinal parasites eggs arc present in sewage sludges, and a significant part of them are disease-causing agents. The amount of pathogens found in the sludge is inversely proportional to the sanitary conditions of the community. Therefore, the greater the prevalence of water-borne diseases in the community, the greater will be the required care to handle the sludge, mainly when the disposal route is farming recycling.

The degree of sludge pathogenicity can be substantially reduced through stabilisation processes, such as aerobic or anaerobic digestion, as detailed in Chapter 4. However, many intestinal parasites, and mainly their eggs, are scarcely affected by conventional stabilisation processes, needing a complementary stage or even further stabilisation to achieve complete inactivation. These processes are known as PFRP (Processes to Further Reduce Pathogens).

It is important to point out that what is meant is not a complete disinfection process, since not all pathogenic organisms present in the sludge are thoroughly inactivated. The aim is to reduce the pathogenicity of the sludge to levels that will not cause health risks to the population, according to the requirements for each sludge use.

ISBN: 1 84339 166 X. Published by IWA Publishing, London, UK.

6.2 GENERAL PRINCIPLES

6.2.1 Objectives of pathogen reduction in the sludge

Pathogen reduction in the sludge is introduced into the wastewater treatment plant to assure a sufficiently low level of pathogenicity to minimise health risks to the population and to the workers that handle it, and also to reduce negative environmental impacts when applied to the soil. Therefore, the need for any complementary pathogen removal system will depend on the characteristics of the selected final disposal alternative.

Sludge application in parks and gardens with public access, or its recycling in agriculture has a higher level of sanitary requirements than other disposal alternatives, such as landfills or beneficial use in concrete molds. These requirements can be met by processes to further reduce pathogens and by temporary restrictions of use and public access.

6.2.2 Exposure and contamination hazards

Usually, the diseases are contracted only when human beings or animals are exposed to levels of pathogenic organisms that are sufficient to initiate the infection. The infective dose depends on each organism and on each individual resistance capacity. However, in terms of helminth eggs and protozoan cysts, only one egg or one cyst might be enough to infect the host.

Human exposure to the infective agent may occur through direct or indirect contact, as shown in Figures 6.1 and 6.2.

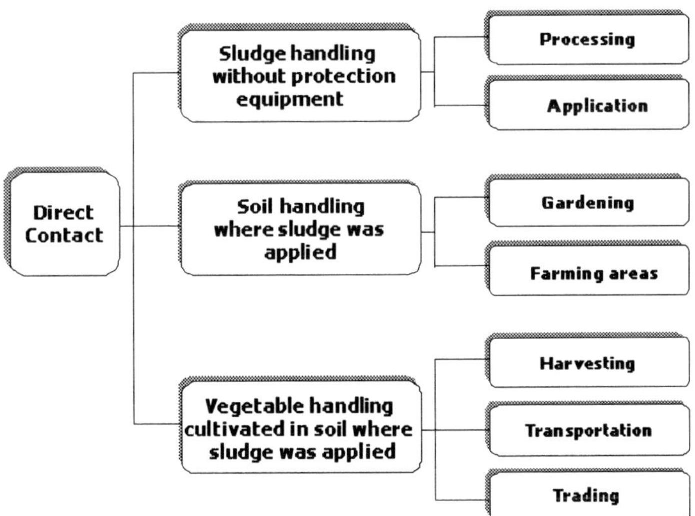

Figure 6.1. Exposure by direct contact

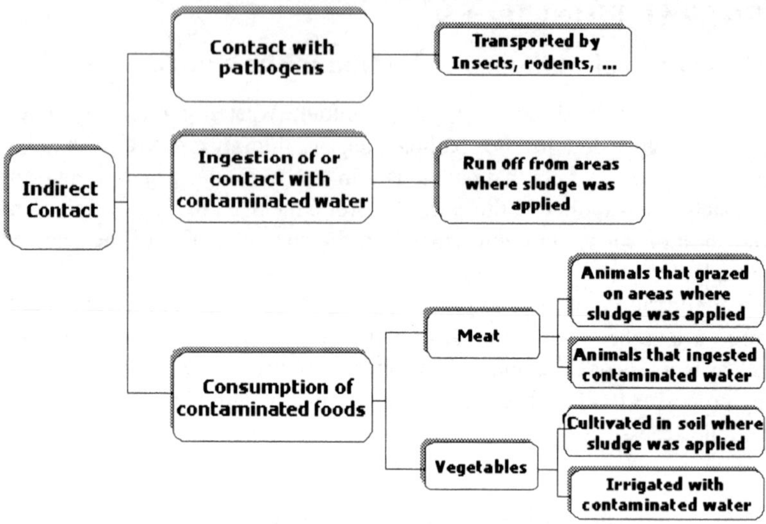

Figure 6.2. Exposure by indirect contact

When sludge is applied onto the soil, environmental conditions significantly affect the survival of pathogenic organisms. A number of organisms promptly die when in contact with hot and dry soils, but they are able to survive for long periods in wet and cold soils. Soils with low pH, organic matter and sunshine exposure (especially ultraviolet radiation) also contribute to inactivation of these organisms.

Bacteria and protozoans are not suitable as public health protection indicators because they are rapidly inactivated by environmental conditions, such as temperature and pH.

As helminth eggs have long survivability, they may be considered to be the most important indicator regarding sanitary conditions of the sludge. However, due to their large size, they usually remain not far from the point where the sludge was land applied.

Viruses, helminths and protozoans are unable to reproduce themselves out of their specific host and do not re-grow once inactivated. However, some bacteria can re-grow when suitable environmental conditions are restored, demanding extra care in any pathogen removal system.

The health of the population and animals can be protected against the potential risk of contamination by the sludge pathogenic organisms through any of the following ways:

- reduction in sludge pathogenic organisms concentration through stabilisation or processes to further reduce pathogens
- reduction of sludge pathogenic organisms transportation by vectors such as insects, rodents, birds etc., through the decrease of the sludge attractiveness to those carriers

- public access restriction to areas where sludge has been applied, for the period of time required for its natural inactivation

6.2.3 Sludge uses and requirements in different countries

Farming recycling is one of the most important and promising sludge disposal routes in most countries. Differently from others though, this alternative requires extra sanitary care, which is dealt with in many countries through use restrictions and/or sanitary requirements.

The technologies available for sludge pathogens removal seek to minimise health hazards through reduction of pathogenic organisms concentrations down to values that allow the unrestricted farming use of the sludge. In general, the limiting values adopted by many countries are very similar, with differences in the approach for use restrictions and in some process parameters.

The European Community criteria (86/278/EEC) require sludge treatment by any biological, chemical, thermal or storage process that significantly reduces the health risks resulting from sludge application to land, and allow each state member to specify its own limits to reach this general goal. This principle seems not to have worked satisfactorily, leading to a review by the European Commission, introducing clearer criteria for the state members to adopt, at least, a minimum common limiting value (Hall, 1998).

Differently from the European interpretation, the United States Environmental Protection Agency (USEPA) adopted, as a control standard and security assurance for public health, two classes of sludge microbiological quality (40 CFR Part 503). **Class-A sludges** have unrestricted use, being produced through processes that assure a concentration of organisms below detection limits, that is, sludges that underwent specific pathogen removal stages. **Classes-B sludges** are those from conventional stabilisation processes and must comply with some constraints and recommendations prior to land application.

South Africa follows a similar criterion, with the sludge being classified into four types, where **Type-C** and **D** can be used unrestrictedly in agriculture as far as pathogenicity is concerned, because the sludge has undergone proper pathogen removal processes.

Table 6.1 compares several pathogen concentration limits in various countries, aiming at achieving a safe sludge for unrestricted farming utilisation. It may be seen that the degree of stringiness varies from country to country.

6.3 MECHANISMS TO REDUCE PATHOGENS

6.3.1 Introduction

Most countries that have legislation for agriculture use of the sludge specify suitable technologies to reduce bacteria, enteroviruses and viable helminth eggs to safe

Table 6.1. Limits of pathogenic organisms concentrations in different countries

Organism	USA (40 CFR 503)	South Africa	European community (86/287/EEC)		European community (new proposal)
			France	U.K.	
Faecal coliforms	<1,000 MPN/g TS	<1,000 MPN/10g TS		Defines only sludge treatment processes and temporary restrictions for planting, harvesting and pasture	Reduction of 6 logarithmic units
Salmonella	<3 MPN/4g TS	0 MPN/10g TS	<8 MPN/10g TS		0 MPN/50g (wet mass)
Enterovirus	<1 MPN/4g TS		<3 MPN/10g TS		
Viable helminth eggs	<1 viable egg/4g TS	0 viable eggs/10g TS	<3 viable eggs/10g TS		

(1) All units on a dry basis, except when otherwise stated

Table 6.2. Time-temperature regimes for Class-A sludges

Regime	Application	Requirements
1	Sludge with at least 7% solids (except those ones secured by regime 2)	Sludge temperature must be higher than 50 °C for at least 20 minutes (0.0139 day)
2	Sludge with at least 7% solids structured as cake, heated by contact with either warm gas or immiscible liquid	Sludge temperature must be higher than 50 °C for at least 15 seconds (0.00017 day)
3	Sludge with less than 7% solids	Sludge must be heated for at least 15 seconds (0.00017 day) but less than 30 minutes (0.021 day)
4	Sludge with less than 7% solids	Sludge temperature must be higher than 50 °C for at least 30 minutes (0.021 day)

Source: EPA (1992)

levels for unrestricted use of the biosolid. Pathogen inactivation is achieved through processes combining thermal, chemical and/or biological mechanisms.

6.3.2 Thermal treatment

Pathogenic organism reduction by thermal route combines two variables: sludge detention time and temperature. Since the sludge has different thermal diffusivities depending upon its solids concentration, USEPA proposes four different time-temperature regimes, that take into account the way the heat contacts the sludge mass, the sludge solids content, the ease of mixing the sludge and the heat transfer capacity. Table 6.2 presents the application and requirements for these four regimes for Class-A sludges.

Figure 6.3 shows the time-temperature relationship for each regime. As it is more difficult to transfer heat for more concentrated sludges (regimes 1 and 2), more conservative relationships are required. On the other hand, considering the lack of thoroughly reliable information, the same relationship is used for sludges with low solids concentrations and contact times shorter than 30 minutes (regime 3).

6.3.3 Chemical treatment

Alkaline products used for pathogen removal raise the sludge pH, consequently lethally altering the colloidal nature of the pathogenic organisms cell protoplasm, and creating an inhospitable environment.

Temperature rising can also take place simultaneously with pH increase, depending upon which product is used. This improves the effectiveness of pathogenic organisms inactivation and optimises the time-temperature relationship

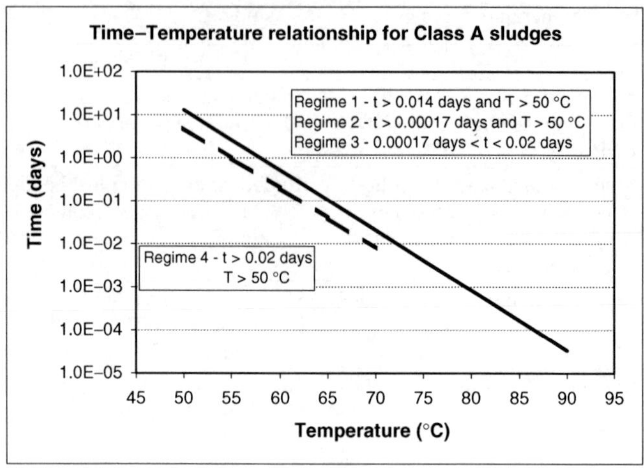

Figure 6.3. Time–Temperature relationship for Class-A sludges (t = detention time; T = temperature)

requirements. A hygienically safe sludge through this mechanism follows the steps shown below.

- Raise the sludge pH to values higher than 12 for at least 72 hours
- Maintain the sludge temperature higher than 52 °C for at least 12 hours, while pH is higher than 12
- Allow open air drying until reaching 50% solids concentration, after the pH-rising period

6.3.4 Biological treatment

The biological route for inactivation of sludge pathogenic organisms still requires further experimentation and more data consistency that would assure reproducibility and scientific acceptance. One of the most well-known alternatives is vermiculture.

Vermiculture is a process in which organic wastes are ingested by a variety of detritivorous earthworms (*Eudrilus eugeniæ*, *Eisenia fetida* and others) and then excreted, producing a humus of great agronomic value that is easily assimilated by plants. When ingesting organic matter, earthworms also ingest pathogenic organisms present in the sludge, inactivating them because of their gastric activity.

However, the presence of gases like ammonia, hydrogen sulphide and carbon dioxide renders the sludge toxic for earthworms, causing their death. In spite of this, there are large-scale plants in Australia and United States, at the present time, which work with a mixture of sewage sludge and other organic wastes, reaching capacities over 400 m^3 per week.

6.3.5 Treatment by radiation

Beta and gamma rays can be used to inactivate pathogenic organisms due to their action on the cell colloidal structures.

Beta rays are formed by electron accelerators under an electric field of one million volt. Their effectiveness in reducing sludge pathogenic agents depends upon the applied radiation dose. As such radiation is unable to penetrate deep through the sludge mass, its effectiveness requires that application be applied through a thin layer of liquid sludge.

Gamma rays are photons produced by radioactive elements like cobalt-60 and cesium-137. As such rays easily penetrate the sludge, this technology can be used in piped liquid flowing sludge, or even dewatered sludge cakes while being transported by belt conveyors. As EPA (1992) recommends, either way requires a minimum one-megarad dose at room temperature for effective reduction of bacteria, enteroviruses and helminth eggs to values below detectable limits. The organic matter present in the sludge is not affected by radiation, so re-growth of pathogenic organisms may occur in case of the sludge being infected again.

Solar radiation, more specifically ultraviolet rays, is well-known by its bactericidal capability. Many researchers have reported inactivation of pathogenic organisms when sludge is exposed to solar radiation. Nevertheless, very little consistent information is presently available about this issue, and whether or not it would be possible to accomplish pathogen reduction to lower the detection level thresholds.

6.4 PROCESSES TO REDUCE PATHOGENS

6.4.1 Introduction

Some processes used for the stabilisation of the organic matter in the sludge are also able to reduce, concomitantly, pathogenic organisms to allow safe use of the sludge. Some specific processes reduce pathogenic organisms to levels lower than detection thresholds and are designated as PFRP (Processes to Further Reduce Pathogens) by USEPA. The most important ones are herein described.

Sludge processing technologies for allowing unrestricted application in agriculture are somewhat similar among the various countries. However, process control variables may differ, reflecting the great variability in environmental conditions and sludge characteristics from one place to another.

Some processes discussed in this section are also covered in Chapter 4 (*Sludge Stabilisation*), since they can be applied for both stabilisation and pathogen removal.

6.4.2 Composting

6.4.2.1 General characteristics

Composting is, in most applications with sludge, an aerobic decomposition process of organic matter achieved through controlled conditions of temperature,

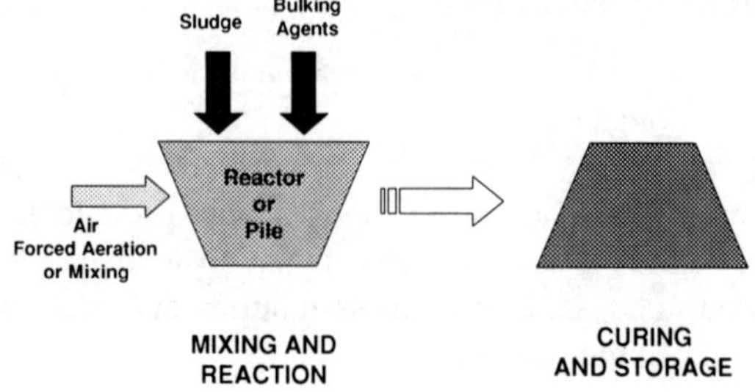

Figure 6.4. Composting process flowsheet

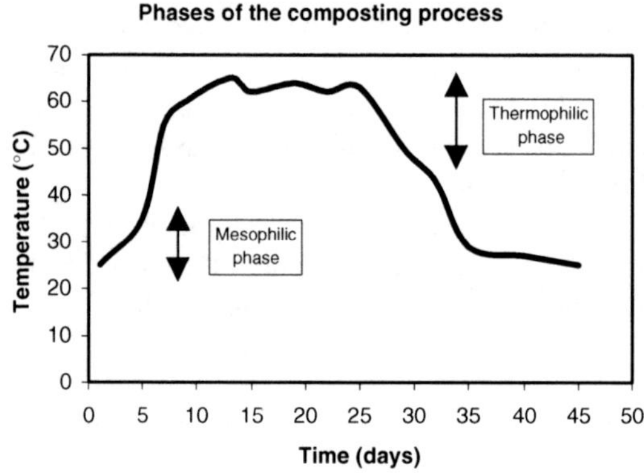

Figure 6.5. Stages of composting process

moisture, oxygen and nutrients. The resulting product from this process has great agronomic value as a soil conditioner. The inactivation of pathogenic organisms takes place mainly via thermal mechanism, brought about by the temperature rise when maximum microorganisms activity is occurring.

Both raw and digested sludge can be composted. Materials such as woodchips, leaves, green residues, rice straw, sawdust or other bulking agents must be added to the sludge to improve moisture retention, increase porosity and balance the carbon-to-nitrogen ratio.

Figure 6.4 shows a typical composting process flowsheet.

The process takes place in three basic stages, as illustrated in Figure 6.5.

- *Initial mesophilic phase.* Fast mesophilic organism growth takes place, with gradual temperature increase.
- *Thermophilic phase.* The percentage of mesophilic organisms decreases as temperature rises, leading to thermophilic bacteria and fungi growth.

These organisms have high activity and reproduction capacity, causing a further temperature rise, thus inactivating the pathogenic organisms.

- *Final mesophilic phase*. As organic matter is exhausted, the temperature lowers and the thermophilic bacteria population decreases, which enables mesophilic bacteria to establish themselves again (although with less activity, as a result of organic matter shortage).

6.4.2.2 Control parameters and environmental requirements

The main environmental requirements and control parameters for an efficient composting process are:

(a) ***Carbon/Nitrogen ratio***

Carbon represents the energy source for composting, while nitrogen is necessary for the reproduction of bacteria (protein synthesis). The balance between these two parameters assures the effectiveness of the process.

Ideal C/N ratio for sewage sludge composting should range from **26–31** (Oorschot *et al.*, 2000). If C/N ratio is higher than this, organisms will not find enough nitrogen, have their growth limited, and the process will become slower, not reaching the temperature required for pathogen destruction. If C/N ratio is lower than the above range, nitrogen is lost due to ammonia stripping, decreasing the compost quality (Fernandes, 2000). C/N ratio must range from **10–20** by the end of the process, which is considered adequate for final disposal.

The introduction of other carbon sources helps to raise the C/N ratio, since sludge has usually very low ratios. Table 6.3 presents the carbon and nitrogen contents in the major agents used for composting.

(b) ***Physical structure***

Sewage sludge has a very fine granulometry, which leads to air distribution problems due to lack of void space among particles. Mixing sludge with vegetable

Table 6.3. Characteristics of the major agents used for composting

Agent	% solids	% N	% C
Tree pruning	65–75	0.8–1.2	45–55
Rice straw	80–90	0.9–1.2	35–40
Sugar-cane bagasse	60–80	0.1–0.2	40–50
Wheat straw	80–90	0.3–0.5	40–50
Sawdust	65–80	0.1–0.2	48–55
Raw sludge	1–4	1–5	30–35
Digested sludge	1–3	1–6	22–30
Dry digested sludge (drying beds)	45–70	1–4	22–30
Dewatered digested sludge (belt press)	15–20	1–4	22–30
Dewatered digested sludge (centrifuge)	17–28	1–4	22–30

Source: Adapted from JICA (1993), UEL (1999), Metcalf and Eddy (1991) and Malina (1993a)

wastes, straw, woodchips and others, chopped in 1–4 cm sizes, increases the porosity within the sludge mass. A **30–35%** porosity usually allows adequate aeration. The bulking material should also lead to a satisfactory C/N ratio, as mentioned above.

(c) ***Moisture***

Moisture must be monitored from the beginning to the end of the process, since it directly affects the reaction rates. Ideal water content levels are **50–60%**, with higher values hindering the passage of free air through the empty spaces, leading to anaerobic zones. Moisture values lower than 40% inhibit bacterial activity and temperature rise for pathogenic organisms inactivation.

(d) ***Aeration***

Adequate oxygen supply is essential for the growth of aerobic organisms, which are mainly responsible for the process. Oxygen supply shall be enough to facilitate the reaction rate control and to assure aerobic conditions throughout the mass under composting. These are essential factors for temperature rise and inactivation of pathogenic organisms.

Some systems use natural aeration while others use forced aeration, with direct introduction of air into the mass core. Excessive aeration decreases moisture and reduces pile temperature, causing problems to the final product quality. Forced aeration systems require an accurate estimation of the oxygen needed along all process stages to assure adequate pathogen removal.

Stoichiometrically, the average oxygen demand is 2 kg O_2 per kg of volatile solids. Rates ranging from **12 to 30 m^{3}air/hour per kg of dry mixture** are normally used in the beginning of batch processes. These rates may be increased along the process, reaching up to **190 m^{3}air/hour per kg of dry mixture** (Malina, 1993a; WEF, 1998).

For natural aeration systems, USEPA recommends revolving the mixture at least five times during the thermophilic phase. Continuous and complete-mix systems demand about 43 kg of air per kg of mixture, which is equivalent to **1,200 m^{3}air/hour per dry ton**. Usually 0.5–2 HP blowers are adequate for this volume.

(e) ***Temperature***

Temperature is an easy-to-follow parameter that indicates the equilibrium of the biological process and hence its effectiveness. **During the first 3 days**, a temperature range between **40 °C** and **60 °C** indicates that the process is running adequately. Otherwise, some environmental requirement (C/N ratio, moisture or pH) is probably not being satisfied (Fernandes, 2000). The ideal temperature for the **thermophilic phase** is **55–65 °C**. At higher temperatures, the bacterial activity decreases and the required cycle becomes longer. At lower temperatures, insufficient decrease of pathogenic organisms may occur. The temperature control can be accomplished by increasing aeration, helping to dissipate the mixture heat released through the reaction.

Table 6.4. Temperature and time required for pathogen inactivation in composting

Organism	Exposure time (minutes)					Remark
	50 °C	55 °C	60 °C	65 °C	70 °C	
Salmonella	10,080		2,880			real
Salmonella			30		4	laboratory
Type-1 Poliovirus			60			real
Ascaris lumbricoides			240	60		real
Ascaris eggs	60	7				laboratory
Mycobacteria tuberculosis				20,160	20	real
Escherichia coli			60		5	laboratory
Faecal coliforms					60	laboratory
Entamoeba histolytica	5					laboratory
Necator americanus	50					laboratory
Virus					25	laboratory
Shigella	60					laboratory

Source: Adapted from WPCF (1991), JICA (1993), UEL (1999)

Table 6.4 shows the time required for inactivation of some pathogenic organisms during composting process, at several temperatures. Significant differences can be noticed between full-scale and laboratory-scale operations.

(f) ***pH***

pH is an important parameter for microbial activity. The best range is **6.5–9.0**. pH reduction may happen in the beginning of the composting process, due to organic acids production, but this issue is solved as soon as the process reaches the thermophilic phase. Therefore, if the C/N ratio of the mixture is adequate, the pH will not usually be a critical factor (Fernandes, 2000).

6.4.2.3 Composting methods

The composting process can be accomplished by three main ways:

(a) ***Windrow***. The mixture is placed in long windrows (Figure 6.6), 1.0–1.8 m high, 2.0–5.0 m wide. The windrows are mechanically turned over and mixed at regular intervals, for at least 15 days (EPA, 1994) or until

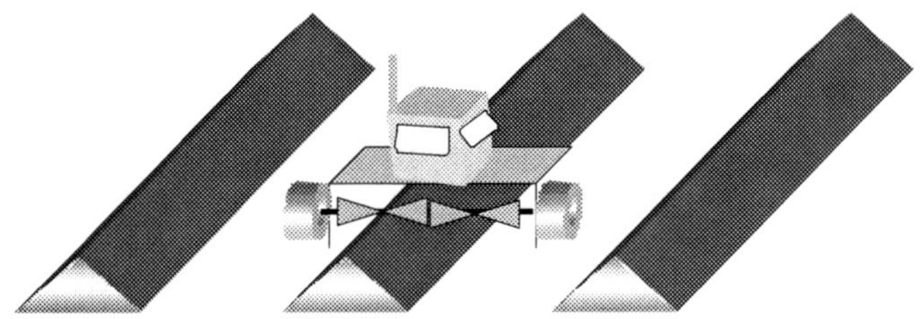

Figure 6.6. Windrows

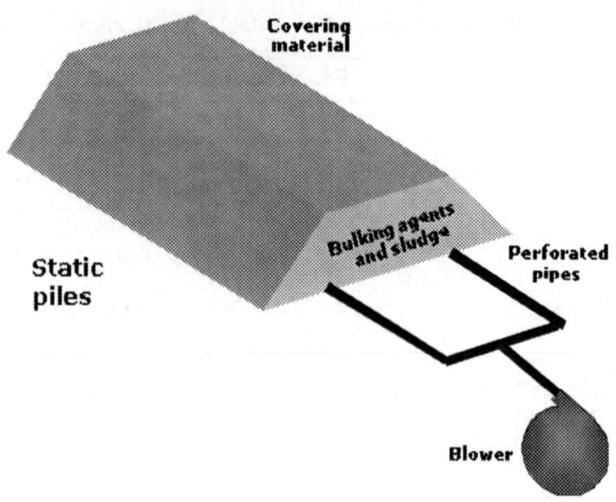

Figure 6.7. Aerated static pile

the process is completed. During this period, the temperature must be kept at least at 55 °C, which is difficult to attain in cold climate countries. The complete process (including the curing time) normally takes 50–90 days up for proper stabilisation. Windrows are usually open-air built, except in heavy rainfall areas. Aeration occurs by natural means through air diffusion into the mixture and by periodical turnover. Land requirements are the highest among the composting processes, being approximately 0.40 ha per 1,000 kg of composted dry solids per day.

(b) ***Aerated static pile***. The mixture is laid over a perforated pipe network, through which air is mechanically blown or aspirated (Figure 6.7). There is no turning over of the pile. The air, after passing through the pile, must be confined and treated to avoid dissemination of foul odours. This treatment can be accomplished by biological filters made up by turf, local soil, fern stems, stabilised compost and other media enabling air filtration through its mass. The thermophilic phase at 55 °C shall be kept for at least 3 days (EPA, 1992). The process is generally completed after 30–60 days, of which 14–21 days are under aeration (WEF, 1998). Land requirements for aerated static piles are about 0.13 ha per 1,000 kg of composted dry solids per day.

(c) ***In-vessel system***. The mixture is enclosed in vessels where all process variables are controlled and odour release is minimum. Shorter reaction times are obtained and better pathogen reduction is accomplished. They can be batch or continuously operated, depending on the project. The process is generally completed after 28–35 days, of which, at least, 14 days inside the vessel (WEF, 1998). Land requirements (around 0.06 ha per 1,000 kg of composted dry solids per day) are much smaller than with other open air processes.

Table 6.5 presents a comparison among the three composting methods.

Table 6.5. Comparison among the composting methods

Composting methods	Advantages	Disadvantages
Windrow	Low investment cost Low O & M cost	Large land requirements Possible odour problems Difficulty in reaching the necessary temperature Potential mixing problems Long composting period
Aerated static pile	Better odour control Better conditions for maintaining temperature Lower reaction time	Investments for the aeration system Moderate O & M costs
In-vessel system	Low land requirements High degree in process control Ease in controlling temperature and odours	Higher investment and O & M costs Economically applicable only for large scale

O & M = operation and maintenance

Example 6.1

Design a windrow composting system for the sludge from Example 2.1 (dewatered in drying beds), using tree pruning as bulking agent.

Solution:

(a) Sludge characteristics

- Sludge production = 4 m^3 per day = 4.2 tonne/day
- Solids concentration = 60% (drying bed dewatered sludge)
- Nitrogen content = 2.5 % (assumed, Table 6.3)
- Carbon content = 25 % (assumed, Table 6.3)

(b) Bulking agent characteristics (Table 6.3)

- Solids concentration = 70%
- Nitrogen content = 1%
- Carbon content = 50%

(c) Required quantities for windrow formation

- *C/N ratio*

C/N ratio must be in the 26–31 range. The C/N ratio for a sludge and tree pruning mixture in equal parts with the above mentioned characteristics will be:

$$\text{Mixture C/N ratio} = (25\% + 50\%)/(2.5\% + 1\%) = 21.4\%$$

Example 6.1 (Continued)

This ratio is too low for composting and needs to be raised by changing the proportion of input material. Making the mixture C/N ratio equal to 30 and adopting a "1" portion of sludge to "Y" pruning parts, the following relationship will stand:

$$30\% = (25\% + 50\% \times Y)/(2.5\% + 1\% \times Y)$$
$$Y = 2.5$$

This means that 1 part of sludge (in weight) should be mixed with 2.5 parts of pruning leftovers. Thus, for 4.2 tonne/d of sludge, 10.5 tonne/d of tree pruning will be required.

As sludge production is continuous, the required pruning amount must be continuously available. Therefore, a careful evaluation of the availability of this material is recommended. Furthermore, it should not be forgotten that the material to be disposed of has increased almost three times its original quantity.

- *Moisture*

The moisture must be set up in the 50%–60% range. Similarly to the C/N ratio calculation, one has:

Mixture moisture = (Sludge moisture × 1 + Pruning moisture × 2.5)/(1 + 2.5)
Mixture moisture = (40% × 1 + 30% × 2.5)/3.5
Mixture moisture = 32.8%

This is low, compared with the recommended values. The following alternatives may be considered:

- Earlier removal of the sludge from the drying beds, since a sludge with a greater moisture content (lower solids content) is needed
- Use of a wetter bulking agent
- Addition of water to the mixture

In the first alternative, the required moisture of the sludge removed from the drying beds needs to be calculated. Assuming "Y" as the sludge moisture:

$$50\% = (1 \times Y + 30\% \times 2.5)/3.5$$
Y = 100% (which means that the liquid sludge instead of the dewatered one should be used)

If a wetter bulking agent is being considered (second alternative), its required moisture level Y is:

$$50\% = (40\% \times 1 + 2.5 \times Y)/3.5$$
$$Y = 54\%$$

Example 6.1 (Continued)

If water is added to the mixture (third alternative), the following amount of water is required:

Original mixture moisture = 32.8% (or 67.2% solids = 672 kg solids/m^3)
Mixture moisture after additional water = 50% (or 500 kg solids/m^3)
Total mass of material = 4.2 tonne sludge + 10.5 tonne pruning
= 14.7 tonne/day
Existing volume of water = (14,700 kg of mixture)/(672 kg/m^3) = 21,875 m^3

Assuming "Y" as the volume of water to be added, one has:

$$500 = 14,700/(21,875 + Y)$$
$$Y = 7,525 \text{ m}^3$$ (say, about one 8 m^3water-truck per day)

(d) Windrow volume

Mass for daily composting = 14.7 tonne/d

Assuming 30% of void space and a specific weight of 1.1 for the mixture:

$$\text{Volume of material} = 14.7 \times 1.3/1.1 = 17.372 \text{ m}^3/\text{d}$$

Assuming a 1.5-m high, 3.0-m wide triangular pile:

$$\text{Pile length} = (17.372 \text{ m}^3/\text{d})/[(3.0 \text{ m} \times 1.5 \text{ m})/2] = 7.7 \text{ m}$$

A 7.7 m × 3.0 m × 1.5 m pile per day shall be built.

(e) Area required

Allowing a 4.0 m lateral circulation around each pile:

$$\text{Pile area} = (7.7 \text{ m} + 2.0 \text{ m}) \times (3.0 \text{ m} + 2.0 \text{ m}) = 48.5 \text{ m}^2$$

Considering 15 days as the required time to complete the composting, the area required for 15 piles is:

$$\text{Composting area} = 15 \times 48.5 \text{ m}^2 = 727 \text{ m}^2$$

An additional area is required to store the product while it is being cured for the next 40 days:

$$\text{Curing area} = 40 \times 48.5 \text{ m}^2 = 1{,}940 \text{ m}^2$$

Assuming 50% of the area for stock room, office, truck loading and transit and others:

$$\text{Total area required} = (1{,}940 \text{ m}^2 + 727 \text{ m}^2) \times 1{,}5 = 4{,}000 \text{ m}^2$$

For the population of 100,000 inhabitants, the per capita land requirement is 4,000 m^2/100,000 inhabitants = 0.04 m^2/inhabitants.

6.4.2.4 *Operational troubleshooting*

Table 6.6 presents an operational troubleshooting guide for sludge composting systems.

6.4.3 Autothermal thermophilic aerobic digestion

6.4.3.1 *Overview*

The autothermal thermophilic aerobic digestion (ATAD) process follows the same principles of conventional aerobic digestion systems, with the difference that it operates in the thermophilic range due to some changes in the conception and operation of the system. ATAD systems are also covered in Section 4.3.4.

In this process, the sludge is usually previously thickened and operates with two aerobic stages, not requiring energy input to raise the temperature. As the reaction volume is smaller, the system is closed and the sludge solids concentration is higher, the heat released from the aerobic reactions warms the sludge. Temperatures may be higher than 50 °C in the first stage and 60 °C in the second stage. The typical heat production is as high as 14,000 kJ/kgO_2, and the oxygen demand reaches **1.42 kg O_2 per kg oxidised VSS**.

Due to the temperature rise, the process can achieve 60% of VSS removal in a relatively short time. Pathogenic organisms are safely reduced to values lower than the detection limits if the sludge is kept in a 55 °C–60 °C temperature range for 10 days (EPA, 1994). However, treatment plants in Germany are designed for 5 to 6 days (2.5 through 3 days per reactor in series), reaching the same results in terms of pathogen reduction (EPA, 1990).

6.4.3.2 *Operational regime*

The reactors operate with daily semi-batches, in accordance with the retention time defined by the project. This operational regime is an important factor for the effective destruction of pathogenic organisms. Once a day the following sequence takes place:

- aeration and mixing for both reactors are turned off
- part of the sludge from reactor-2 is discharged into the sludge holding tank, lowering the reactor water level
- part of the sludge from reactor-1 flows by gravity to reactor 2
- fresh raw sludge is pumped into reactor 1
- aeration and mixing are both turned on again

All this operation takes about 30 minutes, resulting in an average net reaction time of 23.5 hours.

6.4.3.3 *Design considerations*

As oxygen transfer into high solids concentration sludges (4% to 6% of total solids) is difficult, mixing and aeration effectiveness are the major factors governing the

Table 6.6. Operational troubleshooting guide for sludge composting systems

Method	Problem	Cause	Solution
Aerated Static Pile	The pile does not reach 50–60 °C in the first days of operation	Poor mixing of sludge and bulking agent	Check the oxygen. If it is higher than 15%, reduce aeration.
		The pile is too wet	Increase aeration to reduce moisture. As soon it reaches 50–60%, reduce aeration.
		Over aeration	Reduce aeration. If the temperature does not rise after 2 or 3 days, the pile must be remixed.
	The temperature does not remain between 50–60 °C for more than 2 days	Poor mixing of sludge and bulking agent	Keep oxygen within the 5–15% range.
		The pile is too wet	Increase aeration to reduce moisture. As soon it reaches 50–60%, reduce aeration.
		Over aeration	Reduce aeration. If the temperature does not rise after 2 or 3 days, the pile must be remixed.
	Odour in pile	Low volume of applied air	Check the blower. Check if the pipes are clogged.
		Poor mixing of sludge and bulking agent	Raise the volume of air blown to reduce anaerobic condition.
		Non-uniform air distribution	Check for water within air pipes or pipes clogging condition. Verify manifold project.
Windrow	The pile does not reach 50–60 °C in the first days of operation	Poor sludge and bulking agent mixing	Reduce the cycle time between mixings
		The pile is too wet	Protect pile against bad weather and reduce cycle time between mixings
		Over mixing	Increase the cycle time between mixings
	The temperature does not remain between 50–60 °C for more than 2 days	Inadequate sludge mixture with bulking agent	Reduce the cycle time between the mixings
		The pile is too wet	Protect pile against bad weather and reduce cycle time between mixings
		Over mixing	Increase the cycle time between mixings
	Odour in pile	Poor mixing	Reduce the cycle time between the mixings
		Poor mixing of sludge and bulking agent	Reduce the cycle time between the mixings

Table 6.7. Design criteria for ATAD systems

Parameter	Characteristics
Reactor characteristics	Cylindrical reactors, with 0.5–1.0 height/diameter ratio If diffused air is used, increase height/diameter ratio to 2–5
Sludge feeding	4–6% TS (VS > 2.5%)
Detention time	5–10 days
Temperature	Reactor 1: 35–50 °C Reactor 2: 50–65 °C
Air requirements	4 m^3/hour per m^3 of reactor active volume
Power level	85–105 W/m^3 of reactor active volume
Energy required	9–15 kWh per m^3 of sludge
Energy recovered	20–30 kWh per m^3 of sludge

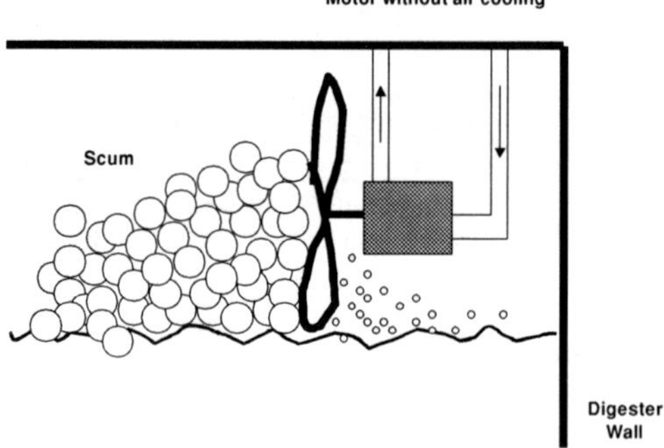

Figure 6.8. Scum and foam controller

operational success. Thus, aerators location, reactors geometry and turbulence conditions are all important design aspects to be considered. Table 6.7 shows typical design criteria for ATAD system.

Oxygenation efficiencies from mechanical aerators operating with 5% total solids concentration sludges are about **1.8 kgO_2/kWh** (standard conditions).

A scum control system must be provided in the reactors, requiring a 0.5–1.0 m freeboard for its installation. Figure 6.8 exemplifies a simple system with a mechanical propeller inside the reactor, which breaks the scum and foam above a certain water level.

The reactor needs to be protected against heat loss by a 10-cm insulation layer.

The sludge holding tank is usually uncovered and equipped with a mixer. Very few problems concerning odour and pathogen re-growth are expected if the sludge is properly stabilised.

6.4.4 Alkaline stabilisation

6.4.4.1 Introduction

Alkaline stabilisation is used for treating primary, secondary or digested sludges, either liquid or dewatered. The process occurs when enough lime is added to the sludge to increase the pH to 12, resulting in a reduction in the percentage of organisms and in the potential occurrence of odours.

6.4.4.2 Liquid sludges

Quicklime (CaO) and hydrated lime [$Ca(OH)_2$] are the most employed products. However, quicklime does not mix easily with liquid sludge and needs to be slaked before application. Hydrated lime is often applied to liquid sludge, which facilitates its mixing, enabling sludge solids and lime to remain suspended in the contact tank.

After a contact time in the mixing tank of about 30 minutes, the sludge is routed for dewatering or immediate land application. As the organic matter is not affected by this process, non-digested sludges must be disposed of before its deterioration starts, avoiding foul odours and minimising risks of pathogenic bacteria re-growth. Figure 6.9 shows a typical flowsheet in which lime is added to a liquid sludge.

The necessary lime dosages for reaching a pH of 12 depend on a number of requirements, such as solids levels, type of sludge, its buffering conditions and others. Table 6.8 suggests some doses as starting values, for later assessment if the mixture has reached the desired pH.

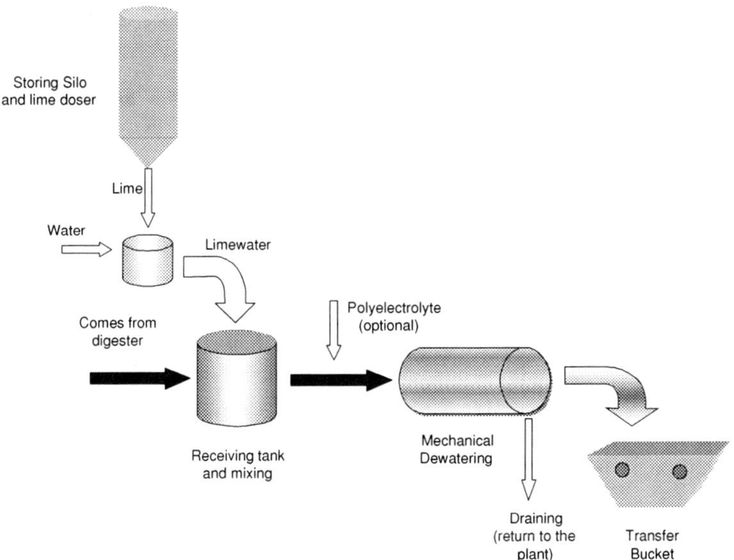

Figure 6.9. Flowsheet of a typical liquid sludge stabilisation system using lime

Table 6.8. Amount of required lime (pH = 12)

Type of sludge	kg $Ca(OH)_2$ per tonne of dry solids	Final pH
Primary sludge	54–154 (110)	12.7
Activated sludge	190–350 (270)	12.6
Anaerobic sludge	125–225 (170)	12.4

Source: Adapted from Malina (1993b)

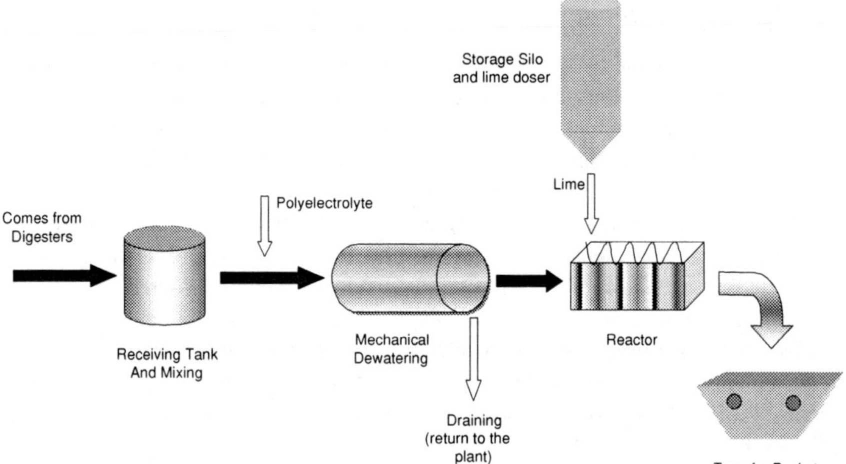

Figure 6.10. Flowsheet of a typical dewatered sludge stabilisation system using lime

The lime-treated liquid sludge is easily dewatered by mechanical equipment, making it suitable for final disposal.

6.4.4.3 Dewatered sludges

Quicklime (CaO) is considered the best product to react with sludges already in the solid phase, since it reacts with the moisture and releases heat. Several studies (Oorschot *et al.*, 2000; Andreoli *et al.*, 1999; EPA, 1992, 1994) have shown that the addition of **30–50% of CaO on a dry weight basis** (0.3 to 0.5 kg CaO per kgTS) to sludge leads to biosolids with pathogenic organisms below the detection threshold. Hydrated lime can also be used, although a significant increase in sludge temperature is not attained, therefore requiring a longer contact time. Figure 6.10 shows a typical flowsheet for the stabilisation of a dewatered sludge using lime.

6.4.4.4 Advantages and disadvantages of alkaline stabilisation

Alkaline stabilisation may present some problems, such as foul odour emission due to ammonia stripping resulting from the pH rise. This is particularly noticeable in anaerobically digested sludges due to the higher ammonia concentration occurring in this process. Although ammonia helps in the removal of pathogens (Sanepar, 1999), the obnoxious odours may pose problems to the workers' health. Attenuation of the problem may be achieved by confinement of the system and gas treatment (scrubbers and air sealed units).

The second problem is related to the increase of solids for disposal. Although in many cases the soil requires pH correction, lime stabilisation leads to a larger amount of solids, increasing transportation and disposal costs.

On the other hand, alkaline stabilisation is an easy and simple technology and does not require high investment or very sophisticated equipment. Therefore, it may be a feasible alternative for small treatment plants or in emergency cases. For other situations, a careful assessment of its economic feasibility is needed.

6.4.4.5 Design and operational aspects for alkaline stabilisation of dewatered sludges

An effective lime and sludge mixing is essential to attain adequate stabilisation and pathogen removal. Furthermore, low-moisture sludges mixed with quicklime have small changes in temperature, losing an important complementary factor for pathogen removal. It is strongly recommended that lime be added when the sludge moisture is at least 60–75%, allowing both an adequate mixing and an exothermic heat releasing reaction from quicklime and the remaining sludge moisture (Sanepar, 1999).

Lime can be mixed with sludge through simplified batch processes (manual, concrete mixer) or continuously by industrial equipments.

- ***Simplified systems***. Manual mixing can be made with spade and hoe. Lime is spread over the sludge surface, while it is still on the drying beds, making a pile out of the two components after mixing. Temperature must be monitored and kept from lowering through pile mixing (Sanepar, 1999). This alternative is somewhat inefficient due to mixing and homogenisation difficulties, both being essential features for an efficient sludge pathogen removal. Another alternative is to use an ordinary concrete mixer, which must be loaded with an adequate mixture of sludge-lime, using about 40% of the active equipment capacity. Mixing takes place for about 3 minutes, followed by unloading for maturation (Sanepar, 1999).
- ***Industrial systems***. Industrial equipments for lime and sludge mixing are produced by several manufacturers. Continuous operation is their major advantage, and they are able to achieve an efficient homogenisation of both components, thanks to ancillary equipment such as lime dosage regulators and mixers.

After mixing, the sludge must remain in a covered place through 60–90 days for completion of the pathogen removal to reduce heat loss and to protect sludge against rainfall. During this period, the pH must be kept around 12 to assure an inhospitable environment for pathogenic organism re-growth.

Sludges that are not adequately stabilised (at least 38% volatile solids reduction) need to be better processed before lime application, because aggressive odours may evolve during storage.

Example 6.2

For the sewage treatment plant defined in Example 2.1, estimate the lime and limed sludge produced to be disposed of.

Data:

- Volumetric sludge production = 40 m^3/d
- Sludge mass production = 1,500 kg SS/d

Solution:

(a) Dewatering

Sludge dewatering before lime addition shall be adjusted for a 25%–40% solids range. Therefore, if drying beds are used, it is important that sludge be removed before the 60% moisture (40% solids) is reached. The following cake production may be expected, assuming centrifuged sludge dewatering (25% solids cake and 98% solids capture):

$$\text{Sludge mass} = 1{,}500 \times 0.98 = 1{,}470 \text{ kg SS/d}$$
$$\text{Cake mass (25\% SS)} = (1{,}470 \text{ kg SS/d})/(0.25) = 5{,}880 \text{ kg cake/d}$$

(b) Lime needed

Assuming 30% lime dose on a dry-weight basis:

$$\text{Amount of lime} = 1{,}470 \text{ kg SS/d} \times 0.3 = 441 \text{ kg/d CaO}$$

(c) Sludge to be disposed of

$$\text{Amount of sludge for disposal} = 5{,}880 \text{ kg/d cake} + 441 \text{ kg/d CaO} = 6{,}321 \text{ kg/d sludge}$$

6.4.4.6 Operational troubleshooting

Table 6.9 presents some common problems, causes and solutions encountered in the alkaline stabilisation process.

6.4.4.7 Other technologies using alkaline agents

Several technologies using variants of lime stabilisation are being offered. Some of them use other additives replacing lime (partially or thoroughly), and reach the

Table 6.9. Common problems in the alkaline stabilisation process

Problem	Causes	Solution
Quicklime is running out of its thermal reaction capability	Air moisture was absorbed during storage, transportation or transfer	Keep storing silo closed and be careful in transportation and transfer
Mixer is locking	Sludge and/or lime feeding is excessive	Adjust dosage
Sludge does not reach the desired temperature	Sludge moisture may be out of optimum range	Adjust sludge moisture
	Quicklime is little reactive	Check lime thermal reaction capability
	The mixture is not adequate	Check mixer
Sludge releases odour after stabilisation	Lime dose was low	Check pH controller and adjust dosage

same removal level of pathogenic organisms. Some examples are:

- *RDP Envessel*: Dewatered sludge is heated before lime addition through a patented equipment that homogenises and heats the mixture, reaching temperatures up to 70 °C. The sludge is then unloaded in furrows and kept for a period not shorter than 15 days.
- *N-Viro*: Dewatered sludge is mixed with quicklime, kiln-dust (a cement industry waste) and an inert product, producing a biosolid with a low odour potential, better granulometry for handling and land application, and safe in terms of pathogenicity.

6.4.5 Pasteurisation

Pasteurisation involves sludge heating up to 70 °C for 30 minutes, followed by a fast cooling down to 4 °C.

The sludge can be heated by heat exchangers or through heated vapour injection. The vapour injection process is more frequently used and the sludge is pasteurised in batches to reduce recontamination risks.

Lately, pasteurisation as a final stage is being gradually discontinued; pre-pasteurisation followed by mesophilic digestion is now the preferred choice, due to some problems concerning *Salmonella* re-growth (EPA, 1993).

Laboratory investigations have proved that pasteurisation of sludges from UASB reactors was able to reduce 100% of faecal coliforms and helminth eggs viability, independently from the solids concentration. However, obnoxious odours evolved after thermal treatment (Passamani and Gonçalves, 2000).

6.4.6 Thermal drying

The application of heat for sludge drying and pathogen removal has been practised for years in several countries. Older technologies, although effective, lacked a good energetic balance. More efficient drying equipment and the growing environmental safety concerns regarding biosolids disposal brought back the debate on this technology again.

In thermal drying, the sludge passes through a heat source that evaporates its water, hence leading to thermal inactivation of organisms. Thermally dried sludge must be previously digested and dewatered up to about 20–35% solids to be economically feasible. Dried sludge has granular appearance and **90–95% solids content**.

Under ideal conditions, 1 kg of water requires 2,595 kJ (0.72 kW) for evaporation, which can be supplied by any heat source, including biogas. As the heating power of biogas is 22 MJ/L and burners can work at 70% efficiency, under ideal conditions, **0.17 litres of biogas are required to evaporate 1 kg of water**. Besides this, energy losses (through walls, air and others) shall also be accounted for, together with the energy required to increase the sludge temperature to slightly above 100 °C, when the evaporation process starts.

The major types of thermal drying systems are:

- *Direct contact dryers:* where hot air has direct contact with the sludge, drawing away moisture, gases and dust
- *Indirect contact dryers:* where heat is transmitted through heat exchange plates

Both systems require equipment for enclosure and treatment of water vapour and dust released from the dryers to avoid odour and particle emissions to the atmosphere.

6.4.7 Other pathogen removal processes

Other processes, such as *incineration* and *wet oxidation*, are operationally more complex and require more capital costs to be implemented. Final products from these processes are inert, sterile, and may serve as concrete aggregate, landfills and alike.

6.4.8 Comparison among processes

A comparison among several process characteristics is shown in Tables 6.10 and 6.11.

6.5 OPERATION AND CONTROL

6.5.1 Operational control

Monitoring the pathogen removal systems aims to ensure a final product meeting microbiological quality requirements, meaning that the concentrations of pathogenic organisms and indicators (*Salmonella* or faecal coliforms, viable helminth eggs and enteroviruses) are below the detection threshold. It should be understood that this is not simply a matter of taking samples, and that the

Table 6.10. Comparison among sludge pathogens removal technologies. Implementation

Process	Area	Skilled personnel	External power	Chemicals	External biomass	Construction cost	O and M cost
Composting (windrow)	+++	+	+/++	+	+++	+	+
Composting (in-vessel)	++	++	++	+	+++	++	++
Autotherm. aerobic digest.	++	++	++	+	+	++	++
Pasteurisation	++	++	+++	+	+	++	++
Lime treatment	++	+/++	+	+++	+	+	++
Thermal drying	+	+++	+++	+	+	+++	+++
Incineration	+	+++	+++	+	+	+++	+++

+++: Significant importance; ++: Moderate importance; +: Little or non-existent importance

Table 6.11. Comparison among sludge pathogens removal technologies. Operation

Process	Effect against pathogens			Product stability	Volume reduction	Odour potential	Remarks
	Bacteria	Viruses	Eggs				
Composting (windrow)	+++/++	++/+	+++/++	+++	↑	+++	Effect depends on mixture
Composting (in-vessel)	+++	+++/++	+++	+++	↑	++	Effect depends on mixture
Autothermal aerobic digestion	+++/++	+++/++	+++	++	++	++	Effect depends on operational regime
Pasteurisation	+++	+++	+++	++	+	++	Must be previously stabilised
Lime treatment	+++/++	+++	+++/++	++/+	↑	+++/++	Effect depends on maintaining pH
Thermal drying	+++	+++	+++	+++	+++	+	Stabilisation and total inactivation
Incineration	+++	+++	+++	+++	+++	+	Stabilisation and total inactivation

+++: Significant importance; ++: Moderate importance; +: Little or non-existent importance
↑: Volume increase

monitoring process should be able to assure that the biosolids beneficial use will not bring forth public health hazards or negative environmental impacts.

Sufficient information must be collected during sludge processing to properly assess whether the system is running as recommended. How information is collected and how frequent sampling must take place varies with the process used.

For instance, the aerated static pile composting process requires that the sludge temperature must be kept at least at 55 °C for more than 3 days, according to USEPA recommendations. Through continuous or intermittent measurements, the operator must ensure that this parameter is adequately maintained throughout the process, and that the measurements really represent the operational conditions of that pile.

Frequently, the process does not behave as predicted. The operator is supposed to record the data of this particular pile for further consideration, when the final product microbiological assessment takes place, to verify whether it should be released or not for disposal.

6.5.2 Requirements for the sludge

To produce a pathogen-free sludge, the operators must assure that concentrations of *Salmonella*, enteroviruses and viable helminth eggs are below the detection thresholds attained by the current analytical methodology.

Faecal coliform and viable helminth eggs have been adopted by some companies (Brazil, Sanepar, 1999) as indicator organisms on the grounds that if their concentrations are kept within the legislation limits the other organisms are also likely to be below the allowable limits. This principle recognises that the long survival capability of helminth eggs under unfavourable environmental conditions should guarantee that a particular biosolid would be suitable for land application.

It is important to mention that the threshold value for the concentration of faecal coliforms represents the geometrical mean of the values found in the samples taken in the monitoring period, defined in Section 6.5.4, and not its arithmetic mean.

6.5.3 Avoiding re-growth

Viruses and helminths, after inactivation, are not capable of appearing again in the sludge, except when external recontamination occurs. On the other hand, pathogenic bacteria may reappear in sludge. Some reasons why this occurs are:

- the sludge is not well stabilised
- the environmental conditions (pH e temperature) which led to pathogen removal are already attenuated
- cross-contamination takes place between the sludge being processed and the final sludge

The conditions that led to pathogen removal must be kept until transportation and final disposal. In addition, any contact between the sludge being processed and the final sludge must be avoided.

6.5.4 Monitoring

6.5.4.1 Requirement and frequency

The monitoring of pathogenic organisms concentration presents two particular problems. The first one concerns the laboratory processing time (about 3–4 days for faecal coliforms and up to 4 weeks for verification of viable helminth eggs). The

Table 6.12. Monitoring frequency

Yearly production (tonne/year)	Frequency
Up to 300	Yearly
From 300 to 1,500	Quarterly
From 1,500 to 15,000	Bimonthly
Above 15,000	Monthly

Source: Adapted from EPA (1994, 1995)

second problem is related to the wide variability of the analytical results, whereby one single sample is unable to assure the sludge sanitary quality.

To cope with this situation it is advisable to adopt **a 2-week monitoring period with approximately seven samplings**. This means that, at least, for 6 weeks, the sanitary quality of the sludge is not known, and hence it should not be routed for final disposal. As a consequence, a storage area is required within the treatment plant, where the sludge shall be duly kept prior to its releasing.

Table 6.12 suggests a minimal monitoring frequency for biosolids that will be land applied. However, the sampling may be intensified in communities with significant records of water-borne diseases.

In the case of yearly frequencies, the operator must choose the most critical period of the year, that is, the one in which the worst performance is expected.

6.5.4.2 Sludge sampling for microbiological analyses

Sewage sludge presumably contains pathogenic organisms, needing to be properly handled to protect operators and laboratory staff. These professionals should be properly trained on sanitary precautions and be given safety equipment, such as gloves, eyeglasses and others, to properly handle the sludge with minimum risks.

All samples for microbiological tests should not be frozen, but rather be preserved at a temperature range of 4–10 °C and processed not later than 24 hours after being collected. Ice and sample shall not have direct contact. All material used for sampling, including the sample containers, must be sterilised (for assessment of compliance with requirements for sludges following pathogen removal).

Samples must be representative of the situation that is being assessed. For instance, thick sludges usually have a concentration gradient, and thus the samples must be collected and mixed from several points throughout the system, both vertically and horizontally. Liquid sludge samples from a tank require discarding the initial volume, which is usually affected by the conditions in the pipe.

6.5.4.3 Before and after measures for sampling for microbiological analysis

The following steps assure reliability of sampling, and therefore represent the results and laboratory safety.

- *Before sampling:*
  - be sure that the laboratory can receive the samples for analysis within the maximum preservation period
  - check whether all equipment and materials that will be used are washed and sterilised (for sludges following pathogen removal)
  - check whether all flasks are identified
  - check whether the system is running normally
- *After sampling and analysis:*
  - assure that all flasks are closed and well packed
  - wash all materials used in sampling before taking them to the laboratory to be sterilised
  - put the remainder unused samples into the autoclave before disposal

7

Assessment of sludge treatment and disposal alternatives

F. Fernandes, D.D. Lopes, C.V. Andreoli, S.M.C.P. da Silva

7.1 INTRODUCTION

The evaluation of alternatives for sewage sludge treatment and final disposal is usually complex, due to the interaction of technical, economical, environmental and legal aspects. Although complex and expensive, final sludge disposal is often neglected in the conception and design of wastewater treatment systems. Operators sometimes need to handle the final disposal of the sludge on an emergency basis, with all the burden of high costs, operational difficulties and undesirable environmental impacts that might undermine the benefits of the wastewater treatment system.

Because sludge management represents a considerable percentage (20–60%) of the operational cost of a wastewater treatment plant, the choice of the sludge processing methods and final destination alternatives should not be overlooked in the design of the wastewater treatment plant, and must be considered a part of the treatment plant itself.

The current chapter presents some basic guidelines that should be taken into account during the assessment of sludge treatment and final disposal alternatives.

ISBN: 1 84339 166 X. Published by IWA Publishing, London, UK.

Existent plants require additional studies regarding land availability, retrofitting, use of already built facilities and/or advantages of building new ones.

7.2 SUSTAINABLE POINT OF VIEW

A hierarchical structure of alternatives is raised automatically when sludge treatment and final disposal alternatives are focused under a sustainable policy point of view. In this case, the following objectives should be attained:

- sludge volume reduction through proper wastewater treatment technology. Although different treatment processes usually have different sludge productions, there is little flexibility to decrease sludge volumes, since usually the higher the wastewater treatment efficiency, the higher the sludge production
- sludge quality improvement through proper management of industrial wastewater in public sewerage systems, with emphasis on their metal content to preserve the possibility of agricultural application of the sludge
- recycling of the produced sludge to the maximum extent. Biosolids land application in crops, pastures and forestry is a worldwide-accepted alternative, together with land reclamation.

From a sustainable point of view, only when the sludge quality makes its beneficial use unfeasible, landfilling or incineration should be considered. Even considering some heat recovery in incineration, its energy balance is negative due to the high water content of the sludge.

Several countries have already adopted economical and legal instruments fostering sludge recycling, and increasing restraints to landfills, deeply influencing decisions regarding the final disposal of wastewater sludges.

7.3 TRENDS IN SLUDGE MANAGEMENT IN SOME COUNTRIES

Sludge production in many countries is dramatically increasing as a consequence of the growth in sewerage and treatment systems. Along with the increase in sludge production, more stringent regulations in terms of a better biosolids quality are gradually being enforced, aiming to minimise adverse sanitary and environmental impacts. These changes are leading to more effective managerial practices, considering the rising trend of final disposal costs.

More mechanical dewatering systems are being used lately because of their improved efficiency in water removal. There is a growing interest in thermal drying, sludge pelletisation and other advanced processes that aim to improve biosolids quality, such as composting, alkaline stabilisation and a number of patented systems.

Many countries have acknowledged that landfill disposal is not a sustainable practice, as they result in greater costs due to transportation over long distances and the increasing environmental restrictions. These factors, associated with the effect of the stimulation policies for recycling, define a clear trend towards using landfills exclusively for non-recyclable wastes. Gains in energy efficacy are being observed in incineration processes, as well as in energy recovery from anaerobic processes and landfills. Incineration is a growing trend in the European Union (EU) and is decreasing in the United States.

Recycling offers the best future perspective worldwide because it is the most economical and environmentally adequate alternative. This final disposal option must be understood as leading to a good amendment for agricultural lands, when used under sound technical orientation to assure a safe environmental and sanitary solution, as well as a cost effective alternative to improve farmers' income. As the quality and environmental requirements become more and more restrictive, there is a trend of increasing costs for such practices.

Environmental restrictions in the EU are greater than those in the United States, especially regarding metals. Biosolids application is quite often limited by its nitrogen content. In the sensitive zones in the EU, allowable nitrogen application rate has been reduced from 210 kgN/ha·year to 170 kgN/ha·year.

The main factors affecting public acceptance of biosolids are related with odour problems during processing and storage. Alternatives for dewatering, stabilisation and advanced biosolids processing methods have shown significant progress lately, as the regulations are enforced aiming at safe biosolid in terms of metal content and sanitary risks. A successful biosolid-recycling programme is a consequence of providing the involved community with adequate information and transparent results on the environmental monitoring programme.

Table 7.1 shows main biosolid management trends in USA and EU countries.

Adequate planning for sludge final disposal determines several characteristics of the plant itself, from its conceptual design influencing sludge quantity and characteristics, up to unit operations as sludge stabilisation, dewatering, pathogen

Table 7.1. Biosolid management trends in the United States and Europe

Processes		United States		Europe
Sludge production	↑	Increasing	↑	Increasing
More efficient dewatering processes	↑	Increasing	↑	Increasing
More advanced techniques for pathogen removal	↑	Increasing	↑	Increasing
Sludge recycling	↑	Increasing	↑	Increasing
Landfill disposal	↓	Decreasing	↓	Decreasing
Incineration	↓	Decreasing	↑	Increasing
Ocean disposal	0	Banned	↓	Decreasing
Legal requirements	↑	Increasing	↑	Increasing
Metal concentrations in biosolid	↓	Decreasing	↓	Decreasing
Power efficiency and energy recovery	↑	Increasing	↑	Increasing
Biosolids management outsourcing	↑	Increasing	↑	Increasing
Biosolids management costs	↑	Increasing	↑	Increasing
Social demands related to environmental conditions	↑	Increasing	↑	Increasing
Farmers' demands regarding biosolids quality	↑	Increasing	↑	Increasing

removal, storage and handling. A number of treatment plants are not equipped with the minimum infrastructure needed for such operations due to inadequate planning and require retrofitting to properly operate the produced sludge.

In many developing countries, demands from society and environmental agencies for better environmental quality are being assimilated by public and private water and sanitation companies. Wastewater treatment plants are gradually being implemented in these countries, therefore causing an increase in the sludge production. Some countries have recently issued land application criteria and now require a feasible sludge disposal plan prior to financing and/or licensing the wastewater treatment plant.

7.4 ASPECTS TO BE CONSIDERED PRIOR TO THE ASSESSMENT OF ALTERNATIVES

7.4.1 Relationship between wastewater treatment and sludge management

As shown in Figure 7.1, there is a narrow link among sewage characteristics, type of treatment and generated sludge.

Some wastewater treatment technologies may either minimise sludge production or produce a sludge that is easier to process. Modern wastewater treatment plants must therefore integrate all the sludge cycle, from its generation up to its final destination, not being restricted exclusively to the liquid phase of the sewage treatment. In addition, sludge management must not be restricted to the solid waste generated by the process, but interact and influence the wastewater treatment system definition.

Before carrying out an assessment on the sludge processing alternatives, the following aspects should be considered.

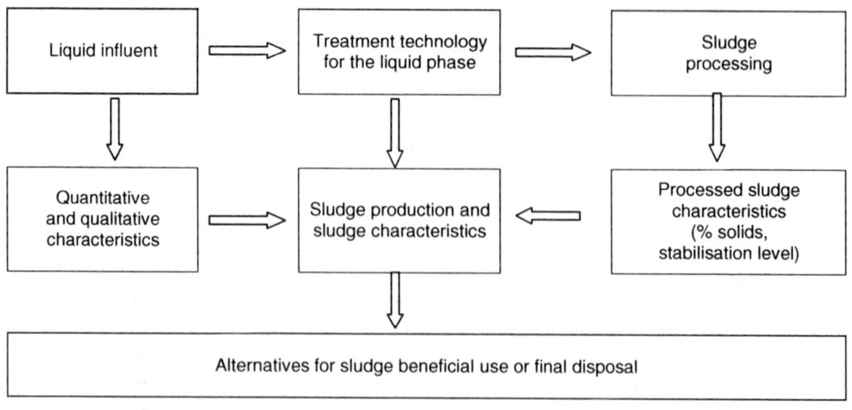

Figure 7.1. Relationship between sewage characteristics, type of treatment, generated sludge and disposal options

7.4.2 Wastewater quality

Metals or organic pollutants contamination may render sewage sludge improper for some uses, such as agricultural recycling. Data regarding sewage contaminants are presented in Chapter 3. Three main possibilities can encompass sewage contamination:

- If the sewage essentially comes from domestic water use, usually no restrictions are imposed towards beneficial sludge uses.
- If the sewage is contaminated with harmful industrial effluents and the sanitation company cannot improve the influent wastewater quality, some sludge beneficial uses (including agricultural recycling) will automatically be discarded if compliance to sludge regulations is not achieved.
- If the sewage is heavily contaminated and the water company is planning to improve its quality, the adaptation to the regulations will demand work with costs and duration varying with the local characteristics.

7.4.3 Wastewater treatment technology

The sludge quality and quantity vary with the wastewater treatment processes as follows:

(a) **Primary sedimentation**. Primary settled sewage sludge is made up of heavier particles with high organic matter content, easily biodegradable and with fast decomposition characteristics, quite often presenting an obnoxious odour. Sludge handling and processing facilities design depend upon sludge volume estimates, which is a function of the characteristics of the raw sewage, including its concentration and age, settling time and characteristics, as well as of the settled solids characteristics, as density and volume, which will depend on tank depth and sludge removal mechanism. The time interval between successive sludge removal operations also influences the sludge volume.

(b) **Chemical precipitation**. Sometimes chemical precipitation is included before or after the biological treatment.

- *Pre-precipitation*. Physical–chemical treatment techniques make up the advanced primary treatment, or chemically enhanced primary treatment. The organic load on the subsequent wastewater treatment stages decreases. However, primary sludge production is increased.
- *Chemical precipitation after biological treatment*. Chemical precipitation after biological treatment improves the quality of the treated effluent, reducing its final organic load and precipitating phosphorus.

Both forms of chemical precipitation significantly increase the sludge volume. Besides, the sludge is unstable and hard to dewater. An increase of 26–35% in sludge mass may be expected when iron or aluminium salts are added to sewage aiming to achieve 1 mg/L of residual phosphorus. Although, theoretically, the sludge increase may be smaller if aluminium sulphate is added instead of ferric chloride, the aluminium precipitate is harder to thicken and dewater due to bonding of water molecules to the formed floc. The phosphorus removed either by aluminium or iron salts is incorporated in the sludge as nearly insoluble phosphates, which is a disadvantage for agricultural recycling. Depending upon the sludge treatment conditions, struvite (Mg, NH_4, PO_4) may be formed and incrustations may happen in filter and belt presses. Use of iron and aluminium coagulants should be evaluated in terms of both sewage treatment and sludge processing and handling facilities.

(c) **Biological treatment**. Biological aerobic or anaerobic wastewater treatment is based on the bacterial metabolism action upon the organic matter. Therefore, biomass is formed, originating in the biological sludge, made up of organic and inorganic constituents (see Chapter 2). Biological sludge may be withdrawn at different time intervals, depending on the treatment process, as detailed in Section 5.3. The discharges may be daily, as in activated sludge systems, or at longer time periods, as in upflow anaerobic sludge blanket (UASB) reactors.

Oxygen in aerobic processes acts as an electron acceptor, and the energy yield is higher, leading to a higher biomass production. Conversely, anaerobic systems are able to convert nearly 90% of the treated COD into methane gas, although this energy does not become available for microbial synthesis. As a general consequence, anaerobic systems produce less excess sludge compared to aerobic ones.

Besides these important differences, anaerobic sludges are more stable and easier to dewater than aerobic sludges.

In both cases, only 25 to 40% of the total biomass produced is biodegradable by natural sludge digestion processes, limiting the possibilities of minimisation of sludge production.

Not only the sludge quantities and characteristics need to be evaluated, but also the operational characteristics of the sludge removal system. For instance, designs of stabilisation ponds generally overlook sludge production, because of their large sludge holding capacity. However, in practice, even with cleaning intervals of many years, the sludge removal is a troublesome and costly operation, difficult to be implemented and variable in performance efficiency.

7.4.4 Scale of sludge production

The total amount of the sludge produced at a treatment plant is an important factor in the definition of the sludge processing technique, mainly from an economic point of view.

Sludge dewatering by mechanical devices, for instance, is usually more advantageous than drying beds from a certain sludge production that allows continuous operation. Mechanised processes for sludge processing and disposal are usually more attractive with higher sludge production values. The economic analysis must define the applicability ranges on a case-by-case basis.

7.4.5 Environmental legislation

All alternatives of biosolids recycling, beneficial sludge uses or sludge disposal must comply with the environmental legislation. Treatment and final disposal of the sludge is an activity that needs to be subjected to environmental licensing, and restricting local conditions must be known before a final decision is made on the processing technology and final sludge destination route.

7.4.6 Soils and regional agriculture

Biosolids can only be land applied in agricultural soil when complying with the pre-requisites towards safety assurance for humans, animals and environment. Of course, land application also depends on an economically feasible transportation distance from the sludge generating points to the disposal sites.

A survey on local crops and the pedological scenario must be undertaken, taking into consideration (see also Chapter 8):

- environmental constraints: nearby water sources, housing developments neighbourhood, conservation areas etc
- pedological constraints: ground slope, soil depth, water table level, soil fertility, hydromorphological quality, top soil texture, rockiness and susceptibility to erosion

In addition to the above preliminary data, pathogen removal techniques and marketing strategies for biosolids distribution should also be duly assessed.

7.5 CRITERION FOR SELECTING SLUDGE TREATMENT AND FINAL DISPOSAL ALTERNATIVES

7.5.1 Relationship between sludge processing and the final destination

Type, size and location of the treatment plant are important issues for the selection of the sludge processing and final destination technologies.

(a) Disposal of liquid sludge. When the wastewater treatment plant is close to agricultural areas and the quantity of sludge produced is not very high, dewatering may be omitted from the sludge processing stage, and the sludge may be applied in the liquid form.

(b) Stabilisation. Sludge stabilisation is very important for agronomic recycling, but has moderate significance to other forms of final destination, such as incineration or landfill disposal.

(c) Conditioning. Sludge conditioning through chemicals addition (coagulants and polyelectrolytes) improves solids capture. The selection of the polyelectrolyte and

Table 7.2. Physical state of sludge as a function of solids content

Total solids content (%)	Water content (%)	Physical state
0–10	90–100	Liquid sludge
10–25	75–90	Pasty sludge • 14–17% of TS: difficulty in storage in high piles • >18% of TS: storage in stable piles with up to 45° slopes
>25	<75	Solid sludge

Source: CEMAGREF (1990)

its dosage will depend upon the type of sludge to be processed. Thermal treatment, besides facilitating sludge dewatering, also may remove pathogenic organisms.

(d) Dewatering. Sludge dewatering has an important impact on freight and final destination costs, and influences sludge handling, because the moisture content affects the mechanical behaviour of the sludge (see Table 7.2). Some alternatives for final disposal require sludge with a well-defined range of solids concentrations. In municipal solid waste landfills in Europe, sludges lower than 15% in solids content usually are not accepted. On the other hand, some dedicated sludge landfills require at least 40% solids to guarantee mechanical stability of the mass. For thermal drying or incineration, at least 35% of solids are required, although the process efficiency increases with higher concentrations. Ideal solids content for composting, depending upon the bulking agent, is in the range of 15%–20%. Sludges with high solids contents, as those dewatered in drying beds, are not suitable for composting due to excessive dryness, which is further increased when low moisture content bulking agents (as sawdust or tree pruning) are added.

(e) Pathogen removal. Pathogen removal is usually needed for agricultural recycling of sludge, because aerobic and anaerobic digestion is unable to keep pathogens below acceptable densities. Pathogen removal processes can be considered as advanced stabilisation processes, since conventional stabilisation, although efficient for the removal of biodegradable organic matter, is usually insufficient for pathogen removal. More recent stabilisation processes such as the Autothermal Thermophilic Aerobic Digestion (ATAD system) are able to eliminate almost all pathogens in the sludge. Composting is a stabilisation process that is capable of producing a sanitarily safe product. For incineration or sludge disposal in landfills, pathogen removal is not really necessary, whereas if agricultural recycling is being considered, lime treatment, composting, thermal drying and others may be used. Each process has its advantages and disadvantages, and only a local analysis can lead to the best alternative.

(f) Criteria for final disposal. Table 7.3 shows the inter-relationship between sludge processing and its beneficial use or final disposal. There should be no fixed rule for the selection of processing and/or final disposal alternatives, but rather a

Table 7.3. Intervening factors on main alternatives for wastewater beneficial use or final disposal

Parameter	Application on the soil	Disposal in landfills	Incineration	Ocean disposal
Sludge treatment				
– Dewatering	+/–	+	+	–
– Stabilisation	+	+/–	–	–
Sludge volume	–	–	+	+
Soil requirements				
– Area availability	+	+	–	–
– Hydrogeology	+	+	–	–
Storage	+	+/–	+/–	+
Good practices	+	+	+	+
Sludge quality				
– Pathogens	+	–	–	–
– Organic pollutants	+	+/–	+/–	+/–
– Metals	+	+/–	+/–	+/–
– Nutrients	+	–	–	–
Public acceptance				
– Odour, aesthetics	+	+	+	+/–
– Traffic	+	+	+/–	–
Transportation	+	+	–	+/–
Energy demand	–	–	+	–

+: Important +/–: Moderate importance –: Without importance
Source: Adapted from EPA, cited by Malina (1993)

judicious study on a case-by-case basis to select the best possible alternative in terms of operational and economical aspects.

7.5.2 Operational performance

The confirmation of the performance of the technical alternative under consideration, in a compatible scale with the case under study, is always an important issue. Foreign technologies must be critically analysed, because they are not always applicable to local conditions or may lead to maintenance problems. Emergent techniques shall also be carefully assessed, considering critical points and fitness to local situations. Sludge processing equipment should run free of problems and its technical and operational simplicity is an important aspect.

7.5.3 Flexibility

Flexibility in an important factor, mainly when changes occur in the quantity or quality of the sludge to be processed. In sludge processing, flexibility is assured when several beneficial uses and final destinations are possible, whereas for final sludge disposal, a flexible solution must be able to absorb fluctuations in sludge quantity and quality. Agricultural recycling may absorb quantity variation, but is

Table 7.4. Relevant items in terms of capital costs

Item	Comment
Required area	Areas needed for buildings, equipment facilities, storage and composting yards. Worksheets are helpful for terrain costs evaluation on the treatment plant site or other place
Equipment	All equipment must be included, such as lime mixers, chemicals dosage equipment, composting equipment, aerators, mixers, thermal dryers and others, as they vary with the requirements of each particular case
Handling material	Pumps, belt conveyors, tractors and trucks necessary to convey the sludge at the treatment plant
Buildings	Foundations for equipment installation, sheds, concrete paved or asphalted areas, laboratories, checkrooms etc
Electric installations	Some sludge treatment equipment calls for special electric installations
Miscellaneous	Experience has shown that estimates should consider a "miscellaneous" item, about 20% out of the total for electromechanical equipment and a value around 10% for civil works

Table 7.5. Relevant items in terms of amortisation and operational costs

Item	Comment
Buildings	Usually 5% (=1/20) as yearly annual rate along 20 years is considered
Electro-mechanical equipment	Usually values for specific cases are defined, based on other systems in operation. • A 14.3% (=1/7) yearly rate over capital costs for a 7-year amortisation period may be assumed in some cases • A 20-year lifetime span has been suggested by some sludge dewatering equipment manufacturers • Well-built limestone-storing silos also may have a 15-to-20-year lifetime • Equipment undergoing quick abrasion and frequently used, such as belt conveyors, containers or pumps, may have a lifetime around 5 years, or a yearly amortisation cost of 20% over capital costs
Maintenance	Widely variable item, depending upon equipment itself and operational care. An average 5% yearly value over acquisition cost may be assumed
Energy	Source of power supply and amount to be used should be known. Volumes and unitary costs should be given for liquid or gaseous fuels. Electric consumption is usually expressed in kWh
Material	Must include all chemicals for sludge dewatering and treatment, as coagulants, lime and bulking agent type and quantities
Handling and transportation	This is a heavy cost impact item, variable with solids concentration and chemicals added
Labour costs	Number of employees, personnel qualification, wages and taxes should be specified
Management and control	Should include physical-chemical and microbiological laboratory analysis costs, office furniture and administrative costs. A 6–9% value over operational costs might be adopted in absence of specific values

Table 7.6. Environmental impacts to be considered in sludge management

Item	Comment
Odours	Relevant regarding both treatment and final destination. May be crucial for agricultural recycling or a secondary factor for incineration
Vector attraction	Closely related to odour, it is a major problem in sludge processing and final destination
Noise	It is an important item in urbanised areas
Transportation	Vehicle and route are the most important features to be considered
Sanitary risks	Although difficult to be objectively evaluated, it may be related to the number of people exposed to sludge handling, sludge quality and infection routes
Air contamination	Air can be contaminated by fumes or particulated matter
Soil and subsoil contamination	Extremely variable issue, depending upon the type and method of sludge final disposal
Surface or underground water contamination	One of the major issues of sludge disposal onto soil or landfill. Risk depends on disposal technique and monitoring control
Increased value or depreciation of nearby areas	Acquisition market value may be objectively evaluated in terms of the surrounding areas
Annoyance to affected population	Some solutions, besides affecting people's lives, may generate resistance groups

limited regarding changes in quality. Conversely, incineration can absorb quality variations, being sensitive to quantity fluctuations.

7.5.4 Costs

Costs are a fundamental issue, and may be split into sludge processing costs, transportation costs and final disposal costs. They can be further divided into capital and operational costs. Because cost estimates are a complex task, it might be of help to group them according to their nature, as shown in Tables 7.4 and 7.5 for capital and running costs. Of course, larger systems demand a higher level of complexity.

7.5.5 Environmental impact

Environmental impacts may be positive or negative. The negative impacts can be minimised through adequate operational procedures. The most relevant impacts are shown in Table 7.6, and are also discussed in Chapter 10.

All factors must be evaluated as a function of local conditions, planned technology to be used and reliability of the monitoring system.

7.6 SLUDGE MANAGEMENT AT THE TREATMENT PLANT

(a) Qualitative and quantitative control

Quality and quantity controls are both essential for technical and financial management of the process. Cost per cubic meter or per tonne of disposed sludge should be considered regarding sludge treatment and final disposal activities. Though handling and freight costs are usually based upon moist sludge (cost per m^3 of sludge), costs per kg TS (dry basis) are preferred when comparing different alternatives. Therefore, a reliable control of the sludge solids level is a relevant issue. The apparent uniformity in sludge cakes from mechanical dewatering systems conceals some variations that may occur in sludge characteristics, chemical conditioning or anomalies due to mechanical malfunctioning. Sludges dewatered in drying beds have variable final moisture. Chemical quality control for agricultural recycling is mainly concerned with metals (Cd, Ni, Pb, Cr, Hg, Zn, Cu), nutrients (N, P, K, Ca, Mg), fixed and volatile solids. The volatile/total solids ratio is a good parameter to assess sludge stabilisation, and hence potential odours generation and vector attracting capability. Minimum biological quality control parameters recommended for agricultural recycling are faecal coliforms (thermotolerant coliforms) and helminth eggs, including viability tests. Sludge quality controls are needed even for alternatives as sanitary landfills or incineration, although with different frequency and parameters.

(b) Sludge handling within the treatment plant

Processed sludge needs to be transported within the plant premises to a proper storage area, or sent to its final destination. If sludge handling occurs inside the plant yard, belt conveyors or trucks are usually employed.

(c) Storage

Storage within the plant premises must be sized based on the average storage time for the predicted sludge volumes and should consider the mechanical characteristics of the sludge. Sludge with solids content in the 12%–15% range has a pasty behaviour and is unable of support itself on 45° slope piles, thus a larger storage area is needed. Handling within the plant yard is done with front-end truck loaders and tipping trucks. The storage area pavement should be drained and impermeable to avoid underground contamination, as well as to facilitate loading and transport operations. If possible, these areas should be covered, avoiding rain from increasing the sludge moisture and minimising odour problems.

(d) Transportation

Suitable transportation depends upon sludge moisture content. While liquid sludges can be pumped or transported by tank-trucks, pasty and solid sludge can be transported by tipping trucks, quite often used in earth-moving works. Truck bucket must be water resistant sealed and latched avoiding drops along the route. A plastic canvas should cover the load during transportation.

(e) Final destination monitoring

Final destination monitoring is a fundamental aspect to assure that objectives of treatment and final destination have been fulfilled. Control parameters and monitoring frequency are function of the type of destination and used technology, usually set by regulations from an environmental protection agency.

(f) Managerial system

The managerial system is the link co-ordinating and evaluating all phases, taking care of all necessary measures towards system efficiency. It must assure that samples are correctly collected and sent to the laboratories, and keep record data well organised and easily accessed to allow checking by state inspection agencies as well as for self-improvement evaluations of the system itself. Operational structure complexity should be compatible with the quantity of treated and disposed sludge. Plant workers might be used in case of small daily sludge volumes production, whereas for large systems a specific crew should be assigned to accomplish such tasks. The managerial system plays a key part in any alternative to wastewater sludge processing and final destination. Even when the water and sanitation company outsources the sludge management, it is still responsible for the safety and efficacy of the process, and must exercise a competent control and supervise all significant activities.

8

Land application of sewage sludge

C.V. Andreoli, E.S. Pegorini, F. Fernandes, H.F. dos Santos

8.1 INTRODUCTION

For thousands of years organic matter has been considered an important soil fertiliser, and organic wastes from human activities were used as fertilisers in ancient times by Chinese, Japanese and Hindus (Kiehl, 1985; Outwater, 1994). In Europe, this became prevalent in 1840 to prevent epidemic outbreaks.

A number of treatment systems during the 19th and 20th centuries have consisted of direct land application of sewage. As technologies for preliminary, primary and secondary biological treatment and chemical precipitation have evolved, sewage land treatment gradually decreased in importance. However, the large increment in sludge production during the 1940s and 1950s, as a consequence of the expansion of sewerage systems, has played an important role in stimulating biosolids recycling whenever possible.

Land application of sewage sludge may be classified into two categories:

- ***beneficial use*: land application of treated sludge (biosolids)**, when advantage is taken on the fertilising and soil conditioning properties of sewage sludge
- ***discard*: final sludge disposal**, when soil is used as a substratum for residue decomposition or storage site, without beneficial reuse of sludge residuals

ISBN: 1 84339 166 X. Published by IWA Publishing, London, UK.

8.2 BENEFICIAL USE

8.2.1 Influence of sludge characteristics on agriculture

From an agronomic point of view, biosolids have nutrients essential to plants, and their presence in the biosolids depend upon the influent sewage quality and wastewater and sludge treatment processes used. Table 8.1 presents biosolid constituents from some wastewater treatment plants in Brazil.

Nitrogen and phosphorus are found in large quantities, whereas Ca and Mg are present in low amounts, except in biosolids treated by alkaline stabilisation. Potassium (K) appears in very low concentration, although in a readily absorbable form by plant roots, being usually supplemented by chemical fertilisers in solids with biosolids addition.

Micro-elements appear in variable quantities in sludges (Table 8.2), usually in higher concentrations for Fe, Cu, Zn and Mn than for B and Mo. When biosolids are applied as the only source of N for plants, the applied quantities of micronutrients often are sufficient for vegetable nutritional needs. It is important to point out that micro-elements are required in small quantities, and if biosolids are applied in larger quantities than crop agricultural needs, toxic effects may occur. Biosolid nutrient concentrations may not supply all plant needs, thus requiring supplemental sources of organic or inorganic fertilisers to cope with the particular crop nutritional needs. The usually supplemented elements are phosphorus – required in large quantities in some soils – and potassium, which has low concentration in biosolids.

Table 8.1. Biosolid constituents from some wastewater treatment plants in Brazil (% of dry matter)

Treatment plant	Type of sludge	N	P	K	Org. C	Ca	Mg	Source
Barueri-SP	Activ. sludge	2.25	1.48	0.01	21.00	7.29		Tsutya (2000)
Franca-SP	Activ. sludge	9.15	1.81	0.35	34.00	2.13		Tsutya (2000)
Belém-PR	Activ. sludge	4.19	3.70	0.36	32.10	1.59	0.60	Sanepar (1997)
UASB-PR	Anaerob. sludge	2.22	0.67	0.95	20.10	0.83	0.30	Sanepar (1997)
Sul-DF	Activ. sludge	5.35	1.70	0.18	34.70	2.68	0.41	Silva *et al.* (2000)
Eldorado-ES	Anaerob. pond	2.00	0.20	0.04				Muller (1998)
Mata da Serra-ES	Facult. pond	2.00	0.20	0.05				Muller (1998)
Valparaíso-ES	Sediment. pond	4.00	3.50	0.07				Muller (1998)

Table 8.2. Micronutrient contents in some Brazilian wastewater biosolids (ppm)

Treatment plant	Type of sludge	B	Fe	Cu	Zn	Mn	Mo	Source
Barueri (SP)	Activated			703	1,345		23	Tsutya (2000)
Franca (SP)	Activated	118	42,224	98	1,868	242	9	Tsutya (2000)
Belém (PR)	Aerobic			439	864			Sanepar (1997)
UASB (PR)	Anaerobic			89	456			Sanepar (1997)
Sul (DF)	Aerobic	22	20,745	186	1,060	143		Silva *et al.* (2000)

(a) Nitrogen

Nitrogen is the element with the highest economic value in biosolids, and to which crops present the highest response. Nitrogen comes from the microbial biomass present in sludge and from residues from the wastewater. In the sludge, nitrogen is present in inorganic (mineralised) forms as nitrates and ammonia, and organic forms as proteins, amino acids, amino sugars, starches, associated with polymers, and others.

The organic fraction makes up most of the N in the sludge, ranging from 70% to 90% depending on the type and age of the biosolid. The mineral forms (nitrite, nitrate and ammonia), although representing a small fraction of the total N, are readily available to plants, whereas the organic N must undergo a mineralisation process, slowly changing into mineral forms readily absorbed by plants.

Nitrogen can only be stored in soil in the organic form. Mineral N is an ephemeral element in soil due to its fast absorption by plants. It may also leak underground or escape to the atmosphere through denitrification (Kiehl, 1985).

Equation 8.1 represents, in a simplified way, the amount of N available to the first crop after the sludge application (adapted from Raij, 1998).

$$N_{available} = f_{org}\,(N_{org}) + f_{vol}(N_{amon}) + N_{nit} \qquad (8.1)$$

where:

$N_{available}$ = N available to the first crop
f_{org} = sludge mineralisation fraction
N_{org} = organic N of the sludge
f_{vol} = 1 – volatilisation fraction of ammonia N in sludge
N_{amon} = ammonia N in the sludge
N_{nit} = nitrate N in the sludge

The organic N mineralisation rate is highly variable, mainly with temperature, moisture and microbial activity in the soil. A general value for the mineralisation fraction (f_{org}) cannot be defined, as it varies widely from place to place and from one year to another. However, it usually ranges from 20% to 70% of the applied organic N. Likewise, the volatilisation fraction of ammonia N is also variable, mainly due to its exposure to the open air. Those losses can be reduced when the biosolid is incorporated into soil, because most of the volatilised ammonia will be trapped by soil particles, remaining available to plants. A typically adopted value of f_{vol} is 30% of volatilisation. Nitrogen in nitrate and nitrite forms, as already mentioned, although readily available to plants, may be quickly washed away by rainfall.

In this way, the sludge may thoroughly meet N requirements of the crops in one single application, slowly releasing the element into soil together with the utilisation by the plants. Literature reports 30% to 50% N availability during the first year of application, decreasing to 10% to 20% in the second year and 5% to 10% in the third year. The remainder quantity is considered part of the humified organic matter in the soil.

The great solubility of N poses a high contamination hazard to groundwater and that is the main reason why biosolid land application for agricultural purposes is normally limited by the N intake crop capability.

(b) Phosphorus

Phosphorus in sludge comes from residues, microorganism cells formed during wastewater treatment and phosphate-containing detergents and soaps. As shown in Table 8.2, sludge is also rich in phosphorus, with a bioavailability of 40% to 80% of the total P.

Plant requirements of P for vegetative growth and production are very low. However, as many soils have high capacity to fix P, the efficiency of chemical fertilisation becomes very low (only 5% to 30% of total P applied through chemical fertilisers are used by plants), which leads to P being the most applied nutrient through chemical fertilisers in many places. Soil may have high amounts of P (100 to 2,500 kg total P/ha), although the assimilable quantity by plants is extremely low, usually 0.1–1.0 kg/ha, due to the high capacity of fixation by solids (precipitation and adsorption).

Optimisation of P use in agriculture through biosolids can be achieved as follows:

- Biosolids can be seen as a P source, assuring a slow and continued release to plants.
- Biosolids influence P cycle in soil, increasing the availability of mineral fixed P, either through acids from organic matter decomposition, which partially solubilises the fixed mineral P in soil, or by chelating the soil soluble P for later release, or still by coating soil components that fix mineral P.

(c) Soil conditioner

Biosolids may still be used as soil conditioner after lime and/or other alkali addition. Such amendment raises pH, reduces toxic levels of Al and Mn, supplies Ca and Mg, improves the absorption of nutrients and stimulates microbial activity. However, caution is required, mainly in saline soils or in soils where Ca + Mg are highly concentrated, due to possible nutritional imbalance, salinisation and pH rising above 7.0, which may hinder crop growth and productivity.

(d) Organic matter

Biosolid organic matter is an excellent soil conditioner, improving its physical, chemical and biological properties, substantially contributing to plant growth and development. Physical characteristics of soil are improved through cementing action, particles aggregation, soil cohesion and plasticity reduction, and increment in its water retention capability. Organic fertilisation generally improves infiltration and retention of water, increasing soil stability of aggregates and its resistance to eroding processes.

The addition of organic matter to a fine-textured soil (silty clay, clay or sandy clay) leads to a change in its structure, increasing its friability and porosity, allowing better air and water circulation and root development. In coarse-textured soils (sandy soils), the addition of organic material aggregates soil particles, thus forming earth clods and allowing the retention of larger volumes of water.

Biosolids can still contribute to improve the soil cation exchange capacity (CEC), a reservoir for plant nutritious elements, as well as pH buffering capacity and the microbial activity within the soil.

Table 8.3 summarises properties of stabilised organic matter and their effects on soils.

Table 8.3. General properties of humus and associated effects on soils

Properties	Characteristics	Effect on soil
Colour	Dark coloured in many soils	Facilitates soil warming
Water retention	Organic matter may hold water up to 20 times its weight	Minimises drought effects and seepage losses
Combination with clay minerals	Cements soil particles and forms aggregates	Facilitates water path through soil and gas exchange, and improves stability of soil structure reducing erosion risks
Chelating ability	Stable complexes may be formed with Mn, Cu, Zn and other cations	Fixes heavy metals and increases availability of micronutrients
Solubility in water	Salts and cations associated with organic matter also become insoluble	Limits organic matter loss due to leaching
Buffering effect	Presents buffering capacity	Helps to keep pH stable in soil
Nutrient holding capacity	Varies from 300 to 1,400 Cmols/kg	Increases soil cation exchange capacity (CEC)
Mineralisation	As the organic matter decomposes, nutrients are released to plants	Source of nutrients
Combination with xenobiotics	Influences bioactivity, persistence and biodegradability of pesticides	Poisonous substances become immobilised, and not absorbed by plants
Energy supply	Contains compounds that supply energy to the micro and mesofauna	Stimulates microbial life, increases soil biodiversity, reduces risks of insects and diseases. Antibiotics and certain phenolic acids are produced, fostering plant resistance to insect and pathogen attacks. Enzymes from microorganisms may solubilise nutrients

(e) Productivity

Better responses of biosolids land application are noticed in degraded soils with prior history of structure and fertility problems. Some experiments in the South of Brazil have shown productivity increases around 32% to 54% (Andreoli *et al.*, 2001). These high increases may reflect the low technological level of the agriculture being practised in the selected areas, since the soil in many cases received no fertilisers. High-technological managed crops present less substantial agronomic responses when biosolids are applied, since productivity was not very low prior to biosolids use. However, it is important to notice the significant economy in chemical fertilisers, mainly in nitrogen, as a consequence of biosolids application, besides the medium-term physical, chemical and biological improvements in the soil.

8.2.2 Environmental aspects

(a) Carbon fixation

The several forms of sludge disposal, be it beneficially applied in agriculture or dumped in a dedicated land disposal site, interfere with the carbon dynamics in our planet.

Carbon on earth is present in several biosphere components, most of it (almost 96%) within oceans and fossil fuels, and only 1.67% in atmosphere. In the last 200 years, anthropic activities considerably reduced biomass carbon, due to deforestation and return to the atmosphere through burning of fossil fuels. From 6 to 8 billion tonnes (Gt) of C released annually into the atmosphere in these processes, 37.5% are transferred to the oceans, and 37.5 to 62.5 Gt are accumulated in the atmosphere (Lashof and Lipar, 1989).

The residence time of CO_2 in soil is approximately 25 to 30 years, and 3 years in the atmosphere. Soils have, therefore, a great potential in the handling of the carbon cycle and, hence, on the greenhouse effect. The estimate for the yearly carbon withdrawal from the atmosphere, considering techniques that increase biomass production in soil, may reach up to 1.23 Gt/year, considering 50% of the soils currently in use.

The influence of the sewage sludge in the retention of organic C in the soil was observed by Melo and Marques (2000) and Melo, Marques and Santiago (1993), with a substantial increase in the organic carbon content and the soil cation exchange capacity. Thus, the agricultural recycling of the sludge, as well as any other form of soil organic matter handling, exploits the direct benefits of productivity increase and the improvement of physical conditions of the soil, and must be stimulated for making up global policies towards carbon cycle balancing.

(b) Erosion and natural resources control

Soil erosion is one of the main environmental problems caused by agriculture, jeopardising the soil productive potential and bringing about large impacts on

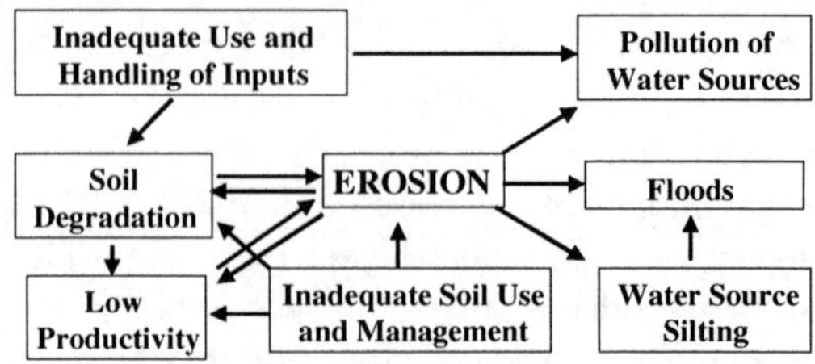

Figure 8.1. Environmental pollution causes and effects in agriculture

the rivers' water quality due to run-off of sediments, nutrients and residues from agricultural toxic products. Figure 8.1 depicts the interaction of the several processes and their interchangeability in terms of cause and effect.

Inadequate use and handling of the soil brings about great damages to soils and rivers. Together with losses due to erosion, most of the agricultural toxic products used in agriculture will end up in the waterbodies. Most of the applied fertilisers are also lost, contributing to the contamination of the water resources.

Strategies for controlling erosion consist in practices for soil handling and use aiming to reduce particles desegregation, increase water infiltration and control surface run-off. The organic matter applied through biosolids furthers a better association of soil particles and improves its structure, stimulates plant root development and the water infiltration through soil layers. It also helps a faster and denser plant growth, quickly covering the top soil, thus reducing raindrop erosion impacts. Table 8.4 lists a number of physical and chemical effects from land application of biosolids and consequences on natural resources conservation.

Table 8.4. Effects of biosolids land use on the control of erosion and environmental pollution

Biosolid	Action in soil	Consequences in the soil	Effects on environment
Organic matter	• Aggregation of soil particles • CEC improvement	• Increases water infiltration • Increases resistance against rainfall impact • Reduces nutrient leaching losses • Improves soil fertility	• Reduces surface run-off • Reduces surface water pollution • Reduces nutrients leaching and groundwater contamination
Nutrients	• Improvement of soil structure • Plant nourishment	• Fosters plant growth • Increases microbial biomass • Accelerates plant growth	• Increases soil covering • Improves soil aggregation

8.2.3 Sanitation sector

Land application of biosolids represents a good alternative for a serious problem that tends to aggravate in many developing countries. The perspective of increase in sewerage and wastewater treatment levels brings about the related increase in sludge volumes.

Land application of biosolids is seen worldwide as a beneficial solution, because it reduces pressure over natural resources exploitation, diminishes the amount of wastes with environmental disposal constraints (Brown, 1993), allows nutrient recycling, improves soil physical structure and leads to a long-term solution for the sludge disposal problem (Andreoli *et al.*, 1994).

Land application is an alternative that associates low cost with positive environmental impacts, when performed within safe criteria. Nevertheless, it depends on adequate planning based on reliable information on wastewater flow and characteristics, suitable agricultural areas at reasonable distances, and managerial capacity to cope with farmers' demand and proper environmental monitoring. Supply of biosolids must ensure a good product for agriculture, also safeguarding public health and environmental aspects.

8.3 REQUIREMENTS AND ASSOCIATED RISKS

8.3.1 Introduction

The main limitations of biosolids land application are soil contamination by metals and pathogenic organisms, and ground and surface water contamination by phosphorus and nitrogen. A worldwide effort has been noticed during the last several years towards the improvement in sludge quality, aiming to lower chemical and biological contamination through better sewer acceptance criteria and improved wastewater treatment technologies.

Chapter 3 addresses the main contaminants in sludge. In the present section, these are focused in terms of the land application of biosolids.

8.3.2 Biosolid quality

8.3.2.1 Metals content

Biosolid metals, when present above certain limits, may be toxic to soil biota, fertilised crops and humans. Since the content of these elements is usually higher in biosolids than in soils, and as their toxicity limits are very low, constant monitoring of the quantities of these elements applied together with the biosolids is required.

The main elements of concern are arsenic (As), cadmium (Cd), copper (Cu), chromium (Cr), mercury (Hg), nickel (Ni), molybdenum (Mo), lead (Pb), selenium (Se), zinc (Zn) and cobalt (Co).

Metal content in biosolids is quite variable (see Section 3.2), depending mainly on the quality of the treated wastewater. Most countries consider heavy metals as a major limitation to land application of biosolids. The available technological alternatives for removal of these elements from biosolids are incipient and very expensive at the present time. The best alternative to preserve biosolid quality is the preventive control of trade effluents and illicit discharges into public sewerage systems.

Monitoring of the metals content in biosolids is the first step in a control programme, even though the most important issue concerns metals accumulation into soil, as a consequence of successive applications of sewage sludge.

Several countries have issued maximum allowable metals concentrations in soil. However, metal dynamics and their toxicity depends on several local factors, such as original content, soil texture, organic matter characteristics, type of clay, rainfall intensity, pH and cation exchange capacity, among others. Care should be taken when defining a generic value for all types of soils, to avoid relaxation or excessive constraints. The ideal is to define values for each large regional geomorphological unit, thus putting together groups of similar soil and climate.

8.3.2.2 Pathogenic organisms

Several disease-causing organisms, including bacteria, viruses, protozoans and helminths, tend to concentrate in the sludge during the treatment process and their quantity reflects the health profile of the sewered population. As pointed out in Section 3.4, they represent a threat to human and animal health, as they can be transmitted through food, surface water, run-off water, and vectors like insects, rodents and birds. To minimise health hazards, biosolids must be submitted to pathogen removal processes to reduce pathogen and indicator densities to values compatible with the applicable legislation and the intended use. Pathogen removal is covered in Chapter 6.

The adopted pathogen removal process affects biosolids management, handling and application. **Alkaline stabilisation** (lime addition), for instance, improves biosolids characteristics, making them suitable to be used as soil amendment. However, the added lime dilutes the nutrients proportionally to the amount added, and leads to a conversion of ammonia nitrogen to free ammonia (NH_3), lost by volatilisation during the maturation period.

Composting may be conducted until the final stabilisation of the biosolid, or be interrupted after the thermophilic phase, in which pathogens have been already eliminated. If undertaken until the final stage, the resulting organic matter will be stabilised and partially converted into humus, producing better effects on soil structure and conditioning. If only the thermophilic phase is reached, the still unripe compost is a good source of nutrients and substrate to the soil biological activity. Similar to limed sludge, it may also act as soil conditioner, after the organic matter has been converted into humus.

Thermal drying is an efficient stabilisation, dewatering and pathogen removal process, involving short-term high temperatures, through which biosolids may reach sterilisation. The resulting material presents low water content, excellent physical aspect (usually granular) and some nutrient loss (mostly nitrogen) from the original biosolid.

8.3.2.3 Organic pollutants

EPA (1979) identifies 114 polluting organic compounds that are discharged into sewerage systems from domestic or industrial effluents. Some are volatised during biological treatment, some are effectively reduced by treatment, while others may reach the sludge processing line and contaminate plant biosolids. Organic pollutants are also covered in Section 3.3.

Some highly toxic and persistent organic micro-pollutants that may be found in sludges are (a) aromatic and phenolic hydrocarbons, (b) pesticides, (c) polybrominated biphenyls (PBB) and (d) polychlorinated biphenyls (PCB).

When applied to land with the sewage sludge, they may decompose through solar energy (photo-oxidation) and undergo volatilisation or biodegradation, which may significantly change their structure or toxicity characteristics.

When directly absorbed by plant roots, some organic micro-pollutants are transported through capillary vessels, reaching the plant's aerial parts.

8.3.2.4 Biological stability and vector attraction

Biosolid storage areas and application sites may suffer from vector insects, small rodents and foul odour release. These problems are a consequence of poor stabilisation and high volatile solids content in the final product, which supposedly should have been eliminated through aerobic or anaerobic sludge digestion treatment stages.

Letting those volatile substances to be eliminated during biosolid storage or after its application poses an open invitation to insects and small rodents, which may trigger the biosolid recontamination process, eventually spreading disease vectors.

There are several parameters allowing the assessment of the degree of stability of the organic matter, and odour emission is one of those. The simplest and more direct methods are the determination of the fixed solids contents (ashes) and the reduction of volatile solids from the sludge.

8.3.3 Risks associated with the biosolids application area

8.3.3.1 Preliminaries

Besides biosolid quality, its safe use depends on suitable environmental characteristics of the site to keep contamination risks as low as possible. Application areas should be selected aiming at the best agronomic results, which will essentially depend on soil aptitude for biosolids application, compliance with environmental constraints and crop restriction.

A suitable soil facilitates biosolid incorporation, fosters biological activity and cycling of nutrients, organic matter and other components, with no hazards to human health, environment and soil potential productivity. Major risks of concern are:

- **groundwater contamination** due to leaching of biosolid components, mainly N, associated with the inner soil drainage
- **surface water contamination** due to surface run-off of biosolid components through soil erosion
- **direct contact of biosolids with humans and animals** due to application in areas close to residences or with public access, and inadequate or absent individual protection equipment, amongst others

The soil productive potential may be jeopardised by physical-chemical and nutritional imbalances, mainly related to soil pH and salt concentration. When the pathogen removal process is accomplished through alkaline agents such as lime, soil pH can rise to inadequate levels, disturbing nutrients availability. The frequent use of limed biosolid plus large amounts of Ca and Mg in the soil may cause nutrient imbalance and even soil salinisation.

8.3.3.2 Soil aptitude

Soil aptitude for biosolid use must be assessed in terms of the soil behaviour regarding its erodibility, inner drainage capability and difficulties it may offer to mechanised equipment.

Good soil characteristics for land application of biosolids are, according to EPA (1979): depth, high infiltration and percolation capacity, fine texture sufficient to allow high water and nutrient retention, good drainability and aeration, pH from alkaline to neutral (to reduce the mobility and solubility of metals).

Some relevant parameters related to soil aptitude to biosolids are shown in Table 8.5.

A soil aptitude rating system for biosolid application, relating the parameters from Table 8.5 with the associated risk level, is presented in Table 8.6 (Souza *et al.*, 1994). The rating system may help to define site aptitude, both at managerial and planning stages, defining preferential application zones from soil maps.

The aptitude itself is obtained from the most restrictive classification. For instance, a particular soil may be in Class I regarding depth, III regarding texture, III regarding its erosion susceptibility, IV regarding topography, I regarding rockiness, hydromorphism and pH. The final aptitude class of this soil will be IV, because higher risks are related to the land steep slope, high erosion and surface run-off.

The aptitude of the areas according to their classes may be interpreted as follows, as far as the potential for biosolids application is concerned:

- **Class-I soils**: very high potential
- **Class-II soils**: high potential

Table 8.5. Parameters involved in the assessment of soil aptitude for the use of biosolids

Parameter	Importance for the definition of aptitude
Depth	As the soil is a good filter, it hampers leaching of sludge components, reducing groundwater contamination. However, high solubility elements, such as nitrogen and potassium, may travel to deep layers and cause problems. Deep soils show lower risks, because of the higher difficulty in transportation and distribution of sludge and its by-products across soil profile. The minimum distance between top soil surface and rock or water table should be 1.5 m
Texture	Soil texture is related to its filtration capability and percolation easiness through soil profile, which may lead to groundwater contamination. Very permeable sandy soils easily leak sludge components. Conversely, very clayey soils hamper drainage
Erosion	Susceptibility to erosion favours transport of sludge components due to surface run-off. Erodibility potential is assessed by soil topography (shape, slope and slope length) and physical characteristics (texture and aggregation)
Topography	Topographical characteristics influence surface water run-off and particle dragging possibilities. Medium-texture soil on flat land poses no risks for sludge application, whereas a sandy-texture soil on slopes higher than 20% will certainly have erosion problems
Water table	Shallow water tables increase the probability of environmental contamination. Larger soil profiles imply longer contact between sludge elements and soil particles, minimising risks of contaminants leaching. A minimum 1.5 m depth from the soil surface to the water table should always be kept
Drainage and hydromorphism	Poorly-drained soils facilitate anaerobic soil conditions and high moisture, both favourable to pathogens survival and harmful to biological organic matter degradation. Hydromorphic soils, a general term for soils developed under conditions of poor drainage in marshes, swamps, seepage areas or flats present very shallow water tables, which may emerge, contaminating water bodies
Slope	Steep areas are susceptible to erosion due to high run-off speed, which may carry the sludge down to lower areas, polluting water bodies. Slopes should not surpass 20%, and recommended values are around 8%
Structure	Structure concerns how soil particles are organised in aggregates, and influences soil water motion, roots penetration and aeration. Difficulties in water infiltration are associated with sludge transport due to erosion, while lack of aeration lowers sludge degradation rate

- **Class-III soils**: moderate potential, strict practices for soil conservation are advised
- **Class-IV soils**: susceptible to be used, provided that compensating criteria like handling and cultivation practices are considered. Risks must be acknowledged if procedures are not strictly obeyed
- **Class-V soils**: under no circumstances should be used, due to unacceptable environmental risks

Table 8.6. Soil aptitude rating system for biosolids application

Factor	Criteria	Degree	Class
Depth (DT)	Ferralsols (oxisols), nitosols (alfissols), deep cambisols, deep inceptisols or deep acrisols/nitosols (ultisols/alfisols)	0–nil	I
	Cambisols (inceptisols) or acrisols/nitosols (ultisols/alfisols) with low-depth	2–moderate	III
	Lithosols (lithic group) or other units with shallow depth	4–strong	V
Surface texture (ST)	Clayey texture (35 to 60% clay)	0–nil	I
	Very clayey texture (>60% clay) and medium texture (15–35% clay) texture	1–light	II
	Silty texture (<35% clay and <15% sand)	2–moderate	III
	Sandy texture (<15% clay)	3–strong	IV
Susceptibility to erosion (SE)	Soils in flat slope (0–3%)	0–nil	I
	Clayey or very clayey soils in 3 to 8% slope	1–light	II
	Medium or silty texture soils in 3 to 8% slope, and clayey and very clayey-texture soils in 8 to 20% slope	2–moderate	III
	Wavy slope soils with sandy and/or abrupt character texture, or 20% to 45% slope associated with very clayey texture	3–strong	IV
	20% to 45% slope with medium and sandy texture	4–very strong	V
	>45% slope or steep slope, independently from its textural class		
Drainage (DR)	Well-drained soils	0–nil	I
	Strongly drained soils	1–light	I
	Moderately drained soils	2–moderate	III
	Imperfectly and excessively drained soils	3–strong	V
	Poorly and very poorly-drained soils	4–very strong	V
Slope (S)	0–3% slope	0–nil	I
	3–8% slope	1–light	II
	8–20% slope	2–moderate	III
	20–45% slope	3–strong	IV
	Higher than 45% slope	4–very strong	V
Rockiness (R)	Soils with no rocky phase	0–nil	I
	Rockiness citation	2–moderate	IV
	Soils with rocky phase	4 –strong	V
Hydromorphic properties (H)	Soils with no indication of hydromorphic properties	0 –nil	I
	Soils with indication of hydromorphic properties	2–moderate	III
	Hydromorphic soils: gleysols (aquatic suborders)	3–strong	
pH	Soils with pH lower than 6.5 for limed sludge applications	0–nil	I
	Any pH-range for composted sludge		
	Soils with pH equal to or higher than 6.5 for limed sludge use	4–strong	V

Source: Adapted from Souza *et al.* (1994)

Example 8.1

Define the aptitude class of the following soil: ferralsol with moderate A horizon, clayey texture, evergreen rainforest, lightly-wavy topography.

Solution:

Using the criteria from Table 8.6, the soil aptitude rating system for biosolid application in the following table shows a soil suitable for land application of biosolids.

Criteria	Remarks	Restriction Level	Aptitude Class
Depth	Ferralsols (oxisols) are deep soils, normally with more than 1.5 m.	0–Nil	Class I
Surface texture	Clayey texture does not represent mechanisation difficulties	0–Nil	Class I
Susceptibility to erosion	The association of clay texture in lightly-wavy relief represents low erosion risks in well-managed soils	1–Light	Class II
Drainage	No draining problems (neither excessive nor poor)	0–Nil	Class I
Topography	Lightly-wavy topography, associated with inadequate handling, may lead to erosion	1–Light	Class I
Rockiness	There is no rockiness citation	0–Nil	Class I
Hydromorphism	It has no hydromorfic properties	0–Nil	Class I
Final classification	II ER1 R1 soil with high potential for biosolids use; suitable handling is recommended to avoid possible erosion		

Example 8.2

A region is being assessed in terms of the soil potential for possible land application of biosolids. The soil types, with the respective areas, are listed below. Rate the soils regarding their aptitude for biosolids application.

- **LRd1: Ferralsol (oxisol)** moderate A horizon with clayey texture rainforest evergreen phase lightly-wavy soil – 64,690 ha
- **LEd3: Ferralsol (oxisol)** moderate A horizon medium texture rainforest subevergreen phase lightly-wavy soil and virtually flat – 19,250 ha
- **LEe1: Ferralsol (oxisol)** moderate A horizon clayey texture rainforest subevergreen phase lightly-wavy soil and virtually flat – 18,630 ha
- **BV(a):** Association **Chernosol** (alfisol) shallow clayey stony texture under-evergreen forest phase strongly-wavy relief + **Lithosol** (Litholic subgroup – Entisols) clayey texture subdeciduous rainforest phase strongly-wavy and hilly topography (basic igneous rock substratum) – 62,610 ha

Example 8.2 (Continued)

- **TRe3: Nitosol** (alfisol) moderate A horizon clayey texture subevergreen rainforest lightly-wavy and wavy soil – 188,250 ha
- **PV3: Acrisol (ultisol)** Moderate A horizon sandy/medium texture rainforest subevergreen phase lightly-wavy soil – 6,560 ha

Solution:

The soil rating, based on criteria from Tables 8.5 and 8.6, is as shown in the following table.

	Limiting factors							
Soil	DT	ST	SE	DR	S	R	H	Class
LEe1	0	0	1	0	1	0	0	II SE1 S1
LEd3	0	1	1	0	1	0	0	II SE1 S1
TRe	0	0	2	0	1 and 2	0	0	II SE1 S1 III SE2 S2
PV3	0	1 and 2	2 and 3	0	1	0	0	IV SE3
LRd1	0	0	1	0	1	0	0	II SE1 S1
BV(a)	4	0	3 and 4	0	3 and 4	3 and 4	0	V

Therefore, the preferential zones map should have four aptitude classes: II SE1 S1, II SE1 S1 + III SE2 S2, IV SE3 and V.

Classes I, II and III soils have extremely high, high and moderate aptitude for biosolids application (see the following table). Only 19.21% of the surveyed area is unsuitable for biosolids (IV and V aptitudes), although fruit growing in Class-IV soil could be practised with special precaution as application in ditches.

Agricultural use of biosolids according to soil classes

Class	Soil	Area (ha)	Percentage of area (%)	Recommended use
II SE1 S1	LRd1	64,690	22.24	Suitable for biosolid application
	LEe1	18,630	5.17	
	LEd2	19,250	6.62	
II SE1 S1 III SE2 S2	TRe3	188,250	64.47	
Total apt area		*290,820*	*80.78*	
IV SE 3	PV3	6,560	1.82	Not recommended
V	RE10	62,610	17.39	Not allowed
Total improper area		*69,170*	*19.21*	
Total area		*359,990*	*100.00*	

Table 8.7. Environmental restrictions and rating of lands for biosolids application

Limiting factor	Minimum distance from application area	Soil class
Vicinity of watercourses, canals, ponds, wells, vegetable-producing plots, residential and public visitation areas	100 m 75 m 50 m	For Class-IV soils For Class-III and II soils For Class-I soils
Water sources for public water supply systems	2,000 m 200 m	Area of direct influence on water source* Area of indirect influence on water source**

* Direct influence: semi-circle area with 2,000-m radius upstream from water abstraction point
** Indirect influence area: up to 20 km upstream from water abstraction point
Source: Andreoli *et al.* (1999)

8.3.3.3 Site selection

Biosolid spreading and incorporation should be properly performed, otherwise it may pile onto soil surface and be carried by rainfall run-offs, concentrating in depressions in the area and eventually reaching watercourses. Even if proper pathogen removal has been undertaken during sludge treatment, inadequate biosolid distribution may alter nutrient and organic matter concentrations in water, leading to pollution and contamination.

Biosolid application sites should not be selected near public access places or housing developments to avoid foul odours, vector attraction annoyances and health hazards.

To keep land application of biosolids feasible, a number of countries have established rules, including restrictions for the areas and the crops that may be raised when using biosolids. The limits can be defined according to site soil aptitude, with less or more stringent criterion depending upon specific site characteristics, as shown in Table 8.7. The concern for potential contamination of public water sources is clearly stated, independently from the soil class.

8.4 HANDLING AND MANAGEMENT

8.4.1 Crops, associated risks and scheduling

Besides biosolid quality and site selection for biosolids land application, good agricultural practices are usually ruled by the local environmental agency, and usually include crops suitability, biosolid allowable application rates, application methods and incorporation alternatives. Important items for safe land application and crops suitability are suggested in Table 8.8.

As cereals usually undergo an industrial process before human consumption, they are the most recommended crops to be grown in sludge-amended soils. They can also feed animals or be incorporated into soil to improve its biological, chemical and physical properties (green manure).

Table 8.8. Indicated species and restrictions for cultivation with biosolids

Item	Specification
Recommendation	• Extensive agriculture, whose products are industrialised or not consumed *in natura* • Reforestation and forest management • Fruit growing, in ditches or incorporated prior to yearly blossoming • Grass, application of the lawn with incorporation • Hazardous land reclamation sites, observing maximum allowable accumulation of metals in soil • Dedicated land disposal, in which all detrimental sludge constituents are kept within the site (least desired option)
Restrictions	• Should not be used for legumes whose harvested parts have contact with the soil • Not recommended for fish culture • The cultivation of legumes and primary contact crops must not occur within 12 months after application of sewage sludge • Pastures: animals grazing should not be allowed within 2 months after application of sewage sludge
Remark	• Sludge submitted to a process to further reduce pathogens (PFRP) can be unrestrictedly used in areas and crops after authorisation granted by the environmental agency

Source: Adapted from Fernandes *et al.* (1999)

Reforestation areas present a special interest, since human consumption is not involved. Also crops such as coffee, sugarcane and tea-crops, which are not eaten raw, represent a potential segment. Metals and excess nitrates reaching groundwater and surface water are usually the limiting factor for a biosolids land application programme.

Fruit growing constitutes a good potential market, due to the high organic matter requirements. High amounts of organic fertilisers are recommended, both in orchard implementation and in yearly holding manuring. Should steep slope hamper the mechanical incorporation of sludge, application may be in ditches during orchard implementation only, keeping biosolid particles from being carried over due to erosion and surface run-off. This restriction must also be observed in coffee plantations.

Higher risk crops are those which have edible plant parts in direct contact with soil (primary contact), mainly legumes and vegetables such as lettuce, cabbage etc., or even below soil surface (carrot, beet, onion, turnip etc.), if they are to be eaten raw.

Animals should not graze on the land for 2 months after application of biosolids as a measure aiming two purposes: (a) to allow a perfect growing of the species in its maximum fodder production and (b) to avoid direct animal contact with the residues.

Nevertheless, if processes for further reduction of pathogens are adopted, these crops may be unrestrictedly fertilised with biosolids.

The above concepts are also relevant for planning purposes. Besides land aptitude, sustained biosolid application still depends on commercial exploitation of the crops in a particular region. The agricultural profile of the focused region is important for both volume assessment of biosolid application and biosolid distribution scheduling all year long, according to the demand of each cropping. A good distribution scheduling has the additional advantage of reducing storage, both in treatment yards and in rural properties. Storage is responsible for many drawbacks, such as odour, insects, physical space requirement, insecure operations and health regulations infringements.

8.4.2 Biosolids application rates

The major agricultural interest in land application of biosolids is associated with its nutrient content, mainly nitrogen, micronutrients and organic matter. Effects of organic matter are felt at long term, increasing soil resistance against erosion, activating microbial life and improving plant resistance against insects and diseases. On the other hand, nutrient effects can be observed at short and medium term. Careful planning is therefore necessary to avoid the application from jeopardising the quality of surface or ground water, as well as the productive potential of the soil.

The control of the application rates, besides being an instrument for controlling fertilisation, is another technical instrument for assessing and controlling the safe use of biosolids. The application rate is a function of the nutrient requirements of the species to be grown, the agronomic quality of the biosolids (mainly N content), the soil of the application site and the biosolids physical–chemical quality (metal content and reactive power).

As the N content in biosolids usually meets cropping needs, application rates are generally calculated as a function of the particular crop nitrogen requirement, whereas P and K are supplemented with chemical fertilisers.

(a) Nutrient recommendation and agronomic quality of the biosolid

The application rate must not lead to an N input greater than the crop requirements to avoid leaching to occur. The amount of N and P in sludge that becomes available through mineralisation of organic matter will depend on previous cropping (nutrients from crop residues) and soil type, and this should be based on a case-by-case assessment. Mineralisation is quicker on sandy soils than on clayey soils. Nitrogen from mineralisation will result in a lower fertiliser requirement. A value of 50% N availability is usually adopted for the first year after biosolid application.

(b) Calcium carbonate equivalence (CCE)

When lime treated sludge is applied, an excessive increase of the soil pH may occur, thus leading to nutrient imbalance.

Neutralisation potential increases with the increase in CCE value, because the acid neutralising potential is associated with calcium carbonate equivalence. One-third of 50% limed sludge (dry basis) is quicklime, which is equivalent

Table 8.9. CCE of some soil amendments used in agriculture

Soil conditioner	Calcium carbonate equivalence (%)
Dolomitic limestone	90–104
Calcitic limestone	75–100
Quicklime	150–175
Slaked lime	120–135
Basic slag	50–70
Gypsum	None

tosaying that when biosolid is land applied at 6 t/ha, a quicklime dose equal to 2 t/ha is also being applied. The quicklime used in the sludge disinfection has a CCE over 150%, whereas limestone, largely used in agriculture, averages a CCE of 75%. Hence, a double amount of dolomitic limestone is needed to achieve the same CCE provided by a given amount of quicklime applied with quicklimed sludge. Table 8.9 presents CCE values of some common soil amendment products.

Most tropical soils have an acidic pH and require liming to increase their productive potential. Limed sludge may effectively substitute limestone application.

Example 8.3

Calculate the biosolid application rate for a medium-productivity corn crop (4,000 to 6,000 kg of grains per hectare). Data:

Soil analysis

pH	Al^{+++}	H+Al	Ca+Mg	Ca	Mg	K	CEC	P	C	m%[1]	V%[2]	Sand	Silt	Clay
	cmolc/dm³							mg/dm³	g/dm³			%		
4.8	0.2	5.7	5.1	3.6	1.5	0.41	14.2	4	19	3.5	38.8	4	20	76

[1] saturation of toxic aluminium: m% = $Al^{3+} \times$ 1,000/(Ca+Mg+K)
[2] saturation of alkalis: V% = Sum of Soil Alkalis/CEC = (Ca+Mg+K) × 100/CEC

Aerobic sewage sludge characteristics (% dry matter)

Type	Total N	Total P_2O_5	K_2O	Ca	Mg	pH	O.M.	Moisture
Raw	5.00	3.70	0.35	1.60	0.60	5.9	69.4	85%
Limed	3.00	1.8	0.20	9.00	4.80	11.4	37.6	80%

Observation: the table shows the composition of both raw sludge and same sludge after lime addition. It should be noticed that the contents of Ca, Mg and pH are different as the limed biosolids keep their concentrations respectively proportional to the applied liming. N losses following the liming process being carried out are also emphasised. In the table, a lime dosage of 50 % of the sludge dry solids is assumed (1 kg of lime per 2 kg of dry solids).

***Example 8.3* (*Continued*)**

Limestone ➔ CCE 75%

Corn plant nutrients requirements

Content in the soil	N	P (mg/dm^3)			K (cmolc/dm^3)		
		0–3	3–6	>6	0–0.15	0.15–0.3	0.3
Productivity (kg grains/ha)		P_2O_5 to be applied (kg/ha)			K_2O to be applied (kg/ha)		
<4,000	50	60	40	30	50	40	30
4,000 to 6,000	80	80	60	40	70	50	40
>6,000	100	90	70	50	110	70	50

Source: Andreoli *et al.* (2001)

Solution:

(*a*) *Crop demand*

From soil analysis and nutrient requirements the following fertilising rates are recommended:

- N — fertilising requirements: **80 kgN/ha**
- P_2O_5: content in the soil: **4,00 mg/dm^3** — fertilising requirements: **60 kgP_2O_5/ha**
- K_2O: content in the soil: **0,41 cmolc/dm^3** — fertilising requirements: **40 kgK_2O/ha**

(*b*) *Amount of available N in biosolid*

$N_{avail} = 0.5 \times N_{bios}$ $\quad$ N_{avail} = Available nitrogen for the first crop

$N_{avail} = 0.5 \times 3.00$ $\quad$ N_{bios} = Total nitrogen in the biosolid

$\mathbf{N_{avail} = 1.50}$ **% dry weight (= 0.015)**

(*c*) *Application rate*

The biosolid dose is calculated as a function of N requirements in the crop (80 kg/ha), dividing this value by the available N content in the biosolid:

$Q_{dry} = R.\ F/N_{\mathbf{avail}}$ $\quad$ Q_{dry} = amount of applicable biosolid (kg/ha)

$Q_{dry} = 80/0.0150$ $\quad$ R.F. = recommendation of N fertilising (kgN/ha)

$\mathbf{Q_{dry} \cong 5{,}300}$ **kg/ha**

(*d*) *Biosolid applied depending on moisture content*

$Q_{moist} = Q_{dry}/(1\text{–}\%\ \text{moisture})$ $\quad$ Q_{moist} = wet sludge (kg /ha)

$Q_{moist} = 5{,}300/(1\text{–}0.8)$ $\quad$ % moisture = moisture (water) content of the sludge

$\mathbf{Q_{moist} = 26{,}500}$ **kg/ha**

Example 8.3 (Continued)

(e) Biosolid effect on pH

To have an estimated effect of limed biosolid on soil pH, total quicklime added together with biosolid must be compared with total lime required to correct soil pH aiming at the planned crop. If the quantity brought by the biosolid is higher, the biosolid application rate should be reduced.

- Checking lime requirement (soil base saturation method – %SB)

$$LR = (V2 - V1) \times CEC \times f/CCE$$

where:

LR = Lime requirement (t/ha)
V2 = SB% (base saturation of CEC) for desired crop production
V1 = SB% of the soil
CEC = Cation Exchange Capacity of the soil
f = incorporation factor (20 cm for biosolids) = 1
CCE = Calcium carbonate equivalent = 120 for quicklime

$$LR = (70 - 38.8) \times 14.2 \times 1/120$$

$$\mathbf{LR \cong 3{,}700\ kg\ of\ quicklime\ per\ hectare}$$

- Quicklime added with biosolid:

$Q_{lime} = Q_{dry}/3$ (biosolid composition: 2/3 sludge and 1/3 lime = lime at 50% dry matter)
$Q_{lime} = 5{,}300/3 \cong 1{,}750$ kg/ha

Soil liming is required.

- Lime supplement required

$$Compl = LR - Q_{lime}$$

where:

Compl = Limestone complement needed (kg/ha)
Compl = 3,700 – 1,750
Q_{dry} = 1,950 kg of limestone/ha

- Availabe limestone required

$$Dose = Compl \times CCE_2/CCE_1$$

where:

Dose = Available limestone dose (kg/ha)
CCE_1 = CCE of limestone in biosolids
CCE_2 = CCE of lime required by soil

Dose = 1,950 × 120/75 = **3,120 kg ha**

Example 8.3 (*Continued*)

(f) Nutrient supplied and required mineral supplementation

Nutrients supplied from land application of biosolid (5,300 tonne dry matter)

Nutrient	Content in biosolid (%)	Available content in biosolid (%)	Application rate (kg/ha)	Recommended rate (kg/ha)	Supplement (kg/ha)
N	3.00	1.50	80	80	0
$P(P_2O_5)$	1.80	0.90	48	60	12
$K(K_2O)$	0.20	0.20	11	40	30

The contribution of the biosolid in the soil fertilising is shown in the graph below.

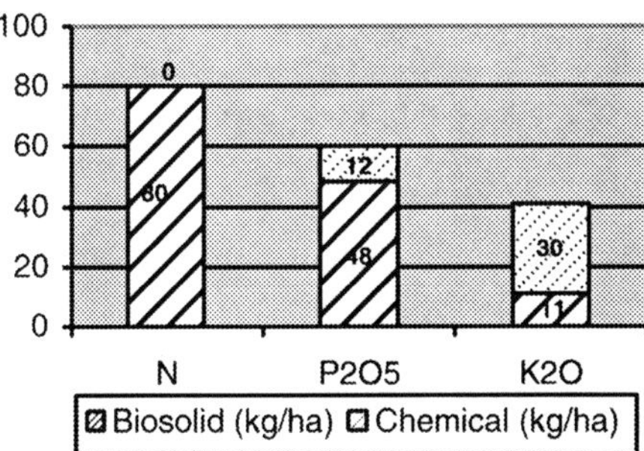

Nutrients in biosolid and in complementary chemical fertiliser

Example 8.4

Check if the biosolid applied dose in the previous example is compatible with the characteristics of the following soil:

Soil analysis

pH	Al^{+++}	H+Al	Ca+Mg	Ca	Mg	K	CEC	P	C	m%	V%	Sand	Silt	Clay
	cmolc/dm^3							mg/dm^3	g/dm^3			%		
5.2	0.2	4.7	5.05	3.2	2.26	0.41	9.1	4	19	3.5	60	30	20	50

Solution:

(a) Calculation of the nutrient supply by the biosolid

From the calculations of the previous example, nutrients were brought by the biosolid at the following doses:

N: 100 kg /ha
P_2O_5: 49 kg/ha
K_2O: 12 kg/ha

All meet crop requirements, but require supplementation.

***Example 8.4* (*Continued*)**

(b) Biosolid effect on pH

- Checking lime requirement (soil base saturation method – %SB)

$$LR = (V_2 - V_1) \times CEC \times f/CCE$$

where:

LR = Lime requirement (t/ha)
V_2 = SB% (base saturation of CEC) for desired crop production
V_1 = SB% of the soil
CEC = Cation Exchange Capacity of the soil
f = incorporation factor (20 cm for biosolids) = 1
CCE = Calcium Carbonate Equivalent = 120 for quicklime

$$LR = (70 - 60) \times 9.1 \times 1/120$$

LR ≅ 0.8 t = 800 kg of quicklime per hectare

- Lime added with biosolid:

$$Q_{lime} = Q_{dry}/3 \text{ (biosolid composition: 2/3 sludge and 1/3 lime} \\ = \text{lime at 50\% dry matter)}$$

$$Q_{lime} = 5{,}300/3 \cong \mathbf{1{,}750\ kg/ha}$$

As $\mathbf{Q_{lime} > LR}$, the application rate shall be corrected.

Soil liming to a pH higher than 6.5 can be detrimental to plant growth, partially because it diminishes essential micronutrients availability.

- Biosolid maximum dose that can be applied:

$$D_{max} = LR \times 3$$

$$D_{max} = 800 \times 3$$

$$\mathbf{D_{max} = 2{,}400\ kg\ of\ sludge\ (dry\ basis)\ per\ hectare}$$

where:

D_{max} = maximum biosolid dose (kg/ha)
LR = Lime requirement (kg of lime /ha)

(c) Nutrient supplied and required mineral supplementation

Nutrients supplied from land application of biosolid (3,900.00 t dry matter)

Nutrient	Content in biosolid (%)	Available content in biosolid (%)	Application rate (kg/ha)	Recommended rate (kg/ha)	Supplement (kg/ha)
N	3.0	1.5	36	80	44
$P(P_2O_5)$	1.8	0.9	22	60	38
$K(K_2O)$	0.2	0.2	5	40	35

The biosolid role as complement of nutrient brought by chemical fertiliser is better visualised in the graph that follows.

Example 8.4 (Continued)

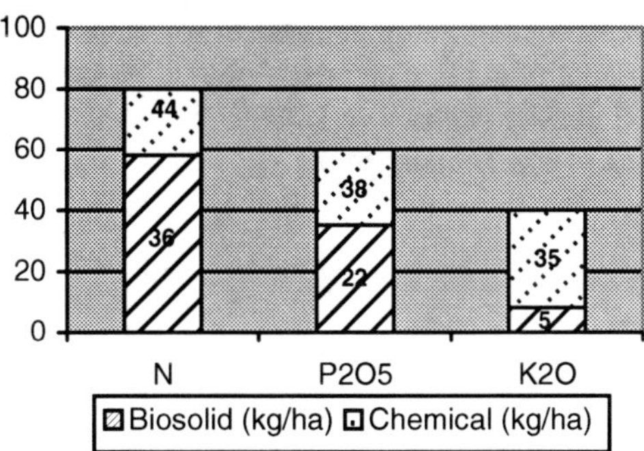

Nutrients in biosolid and in complementary chemical fertiliser

Example 8.5

Evaluate the amount of metal that will be added to the soil from a biosolid land application of 5.55 tonnes (dry matter per hectare), with the following composition:

Sludge characteristics

Element	Content in biosolid (mg/kg – dry basis)
Cd	12
Cu	500
Cr	300
Ni	150
Pb	200
Zn	1,400

Solution:

(a) Added metal quantities

$$Q_{metal} = Bio_{dry} \times C_{metal}$$

where:

Q_{metal} = quantity of applied element (g/ha)
Bio_{dry} = amount of applied biosolid (t dm/ha)
C_{metal} = concentration of the element in the biosolid (mg/kg = g/t)

(b) Metal concentration increase in soil

$$C_{soil} = Q_{metal} \times 1000/(d \times 10{,}000 \times f)$$

***Example 8.5* (*Continued*)**

where:

C_{soil} = Metal concentration increase in soil (mg/kg of soil)
Q_{metal} = amount of applied element (g/ha)
d = soil density (kg/m^3) = 1,200
10,0000 = area of one hectare (m^2)
f = incorporation depth (m) (in the example, 0.2 m)

Thus, the metal quantity brought by the biosolid is:

Heavy metal accumulation in soil due to biosolid application

Element	Total amount applied (g/ha)	Concentration increase in soil (mg/kg)	Yearly maximum rate (g/ha)		Maximum allowable concentration in soil (mg/kg)	
			EU	USEPA	EU	USEPA
Cd	67	0.03	150	1,900	20	20
Cu	2,750	1.15	12,000	75,000	50–210	770
Cr	1,665	0.69		150,000		1,530
Ni	832	0.34	3,000	21,000	30–112	230
Pb	110	0.05	15,000	15,000	50–300	180
Zn	7,700	3.21	30,000	140,000	150–450	1,460

Discussions on maximum allowable concentrations of metals in agricultural soils have been held for years, whenever biosolids land application programmes are considered. Metal concentrations in soil have a wide range variation. Some soils without biosolids amendment have natural metals concentrations higher than the maximum allowable concentrations in several countries. Well managed biosolid land application programmes are demonstrating that no harm, either to harvested crops or to the environment, are expected if legislation limits and sound operational procedures are followed, since the increase in metals concentrations in soil from biosolid amendment is negligible under such conditions.

8.5 STORAGE, TRANSPORTATION AND APPLICATION OF BIOSOLIDS

8.5.1 Storage of biosolids

After the maturation period, that is, the time needed to complete pathogen removal and comply with legal requirements, biosolid is ready to be transported and land applied.

Biosolids may have a continuous or batch production, and differences may occur as far as production and demand are concerned. Biosolids require maturation periods ranging from zero – in thermal drying processes – to 30–60 days in liming processes. As biosolids are under the responsibility of the sanitation company,

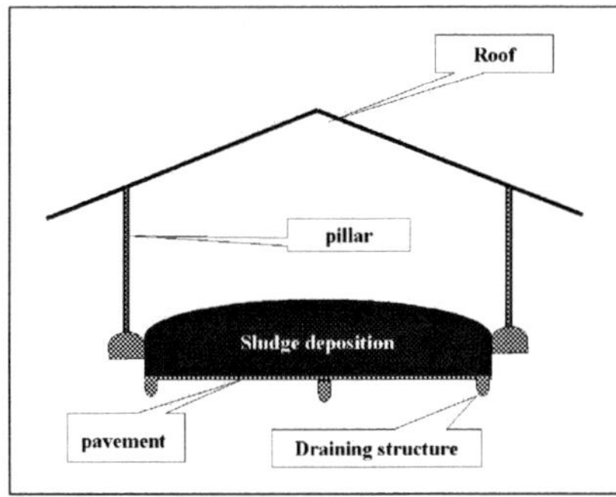

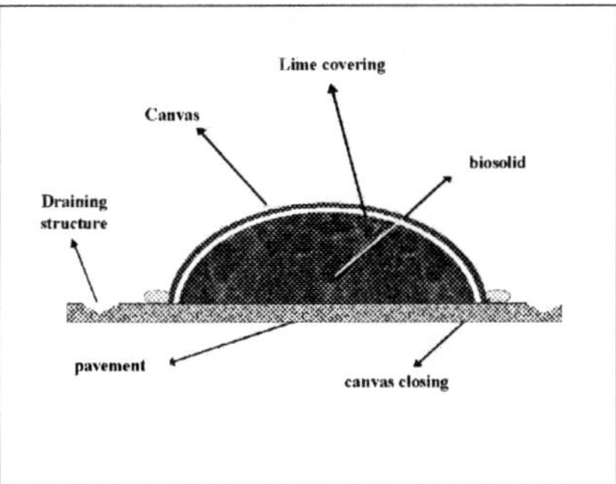

Figure 8.2. Conventional (adapted from Agrodevelopment SA, 1995) and simplified biosolid storage facility

during that period it must remain within the plant premises, and thus a storage yard must be designed and provided for. Figure 8.2 presents a diagram of the basic structure of a conventional and a simplified storage yard.

The storage facility comprises:

- **Paved floor:** pavement is needed to avoid infiltration of leached sludge liquid into the soil. Reinforced concrete or asphalt (pitch) is suitable for that matter. Pitch has higher resistance against chemicals, although less to traffic.
- **Leachate collection network:** leached flow must be redirected to the treatment plant headworks to be treated jointly with the incoming sewage flow, whereas the storm flow (diverted from the storage area perimeter) should join the final effluent disposal.
- **Ceiling height:** no special requirements for canvas covering. It must take into account the operational height of special equipment.
- **Covering:** indispensable for all storage yards. Either a roof or a plastic canvas may be suitable, provided that moisture from rainfall and unauthorised people are both kept away from stored biosolid.

The European Environmental Agency (1997) recommends storage areas of 1.50–0.80 m^3 of biosolid volume per m^2 of storage area for biosolids with mechanical behaviour similar to solids, 0.80–0.40 m^3/m^2 for somewhat plastic biosolids and less than 0.40 m^3/m^2 for wetter ones.

The equilibrium angle increases with the moisture reduction and determines the maximum height of storage heap without side support.

8.5.2 Transportation

Transportation has a major impact on recycling costs and is directly dependent on sludge moisture content. The higher the moisture, the larger will be volumes to

Table 8.10. Amount of biosolids and number of trips necessary for a 6-tonne (dry matter) application

Type of biosolid	Moisture content (average)	Biosolid cake (tonne)	Number of 12 tonne trucks
Liquid sludge	98%	300	25.0
Thickened sludge	92%	75	6.3
Belt pressed	85%	40	3.3
Centrifuged	70%	20	1.7
Filter pressed	60%	15	1.3
Thermal dried	10%	7	0.6

Table 8.11. Biosolids solids content, transportation and handling

Type of biosolid	Typical solids content (%)	Type of transport
Liquid	1 to 10	Gravity or pumped flow, tanker truck transport
Cake	10 to 30	Tipping truck, leak-proof container
Dry pellets	50 to 90	Conveyor, truck

Source: EPA (1993)

be handled and transported, and therefore, the more troublesome and costly the transportation will be.

For instance, if moisture is reduced from 98% to 85%, the sludge volume becomes only 13% of its original volume, as shown in Table 8.10. As the table also indicates, to keep the application rate of 6 tonnes/ha, 15 to 300 tonnes of cake are needed (from dewatered filter-pressed cakes to liquid sludge), whereas only 7 tonnes are required if granulated thermal dried biosolids are used.

Besides sludge volume, also distance, vehicle type, capacity, road conditions and truck loading operation influence transportation costs. Larger carrying capacities lower the unit transportation cost, but roads and traffic shall be compatible, which is not always the case in tropical rural zones. Table 8.11 lists suitable hauling vehicles, depending upon cake moisture range.

Transportation distance and road conditions influence cost items such as fuel, lubricants and maintenance, whereas biosolid characteristics influence the transportation policy (cleaning costs, rejection to product, concerns for contamination). Thus, transportation distance viability is directly influenced by sludge solids content and transportation policy. Table 8.12 recommends safety measures regarding biosolids transportation.

8.5.3 Application and incorporation

Table 8.13 summarises the main biosolid land application practices in agriculture.

Liquid biosolid application is relatively simple. Drying processes are not necessary, and it may be immediately pumped to the application areas. This method

Table 8.12. Necessary cares in biosolids transportation

Item	Precautions
Volume control of hauled material	Bucket volumetric capacity should not be surpassed and freeboard should be kept to the top of the bucket/wagon side structure
Vehicle surface and tires cleaning	Tires and vehicle surface should be thoroughly washed when leaving treatment plant
Cargo covering	Although covering the sludge with canvas may not be necessary for dried sludge with high solids contents, it is a low-cost operation and may avoid undesirable situations
Safety locks	Safety locks should be verified to avoid accidental openings during sludge transportation. Complete check of all container locks should be carried out within plant yard and before truck loading starts
Use of bulk tractor trucks, bulk trucks or other	The sludge container should be leak-proof and perfectly fitted for sludge transportation
Loading and transportation	Loading should not be allowed during rainy days if there is no weather protection

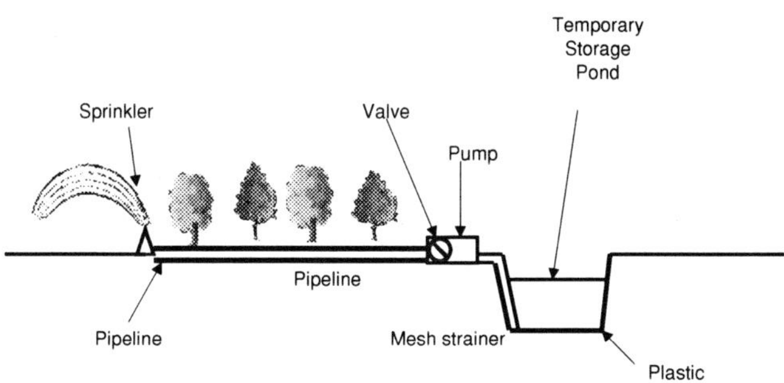

Figure 8.3. Typical sprinkler land application of biosolid (adapted from EPA, 1993)

is generally not used if hauling distance is above 5 km, due to larger liquid sludge volumes involved and implied associated costs. Figure 8.3 shows a sprinkler land application of biosolid.

Dry biosolid may be land applied with equipment used for animal manure application. Depending upon moisture content, pathogen-removed biosolids may vary from pasty (less than 25% solids) to solid (over 50% solids) consistency. Heat dried biosolids with 45 to 65% solids (55 to 35% moisture) have a tendency to stick, as well as 15–35% solids belt-pressed cakes (85–75% moisture), requiring more robust and powerful equipment for spreading.

Heat-dried granular biosolid, typically with 2–4 mm diameter, 90% solids and higher, may be handled, spread and land-applied using classical farming equipment, such as spreaders for lime, damp lime, fertiliser, poultry litter, bedding,

Table 8.13. Summary of main biosolid land application practices in agriculture

Application	Method	Description	Advantages	Disadvantages
Surface application of liquid biosolid	Spray irrigation	Pre-treated liquid biosolids are directly piped and pumped to sprinklers	Traffic in area is minimised Application is feasible on unprepared land	Pond storage or tank storage is needed High storage volume Foul odours during storage and application Sludge adhering to the foliage
	Hauler tank spreading	Liquid biosolid is either injected into soil or sprayed	Haul trucks transport and apply biosolid Smaller applied volumes	Foul odours Site slope limits application Soil may become excessively compacted Depends upon ideal climate and soil conditions
	Farm tractor and wagon spreading	Biosolid is transferred to farming equipment for application	Usual farming equipment	Foul odours during and after spreading Pond storage or tank storage is needed
Surface application of solid or semisolid biosolid	Dewatered sludge haul truck spreader	Solid or semisolid biosolids are transported and spread by trucks	Haul trucks transport and apply biosolid Eliminates odour problems	Solid or semisolid biosolids Unusual equipment in many places Soil may become excessively compacted Depends upon ideal climate and soil conditions
	Tractor powered box spreader	Dewatered biosolid is loaded into the spreading wagon and applied	Usual farming equipment Eliminates odour problems Reduces soil compaction problems	Solid or semisolid biosolid. Depends upon ideal climate and soil conditions
Sub-surface application of liquid biosolid	Tank truck or tractor with a chisel tool	Liquid biosolid is unloaded from tanker and loaded into the vehicle. A subsoil mixing plough is coupled to the vehicle. The biosolid is pumped into the subsoil through hoses coupled to the shafts of the subsoil mixing plough	Minimises odour problems and vector attraction Spreads and incorporates the sludge in a single operation	Unusual equipment in many places Huge volumes to be transported Need for storage pond Site slope limits application

Source: Adapted from U.S. EPA (1993)

compost, gypsum, sand, salt, cement, fly ash and any bulk material. The same equipment is suitable to apply sludge cakes from drying beds after lump breaking of an otherwise sticky sludge (about 50% solids).

Incorporation is a desirable practice and is recommended in a number of biosolid land application regulations. Incorporation avoids people and animals' direct contact with the biosolid and minimises risks of surface water contamination. Biosolid soil incorporation may be performed with classical farming equipment usually found in almost any rural property as disc plough, moldboard plough, disk tiller and chisel plough. Liquid digested or undigested sludge may be soil injected at the pre-determined application rate, 150–300 mm below the ground surface by a purpose-built tractor or truck, which breaks the ground surface and injects the biosolid, resealing the surface afterwards with special press wheels (Santos, 1979).

Regardless of the equipment used, an undesirable soil compaction may occur under the tire tracks which should be dealt with by farmers.

Correct equipment operation is even more relevant than the equipment itself. Ploughs and chisel ploughs promote deeper incorporation of biosolids into the soil, and should not be used unless the soil needs initial preparation. Disc tiller achieves incorporation at 10–15 cm depths, which is sufficient if basic soil preparation was performed before biosolid spreading.

8.6 OPERATIONAL ASPECTS OF BIOSOLID LAND APPLICATION

8.6.1 Introduction

The previous topics dealt with concepts, processes and methodologies required for the beneficial land application of biosolids. This information should be duly assessed and structured to provide a framework for planning activities, organisation, implementation and management of a recycling programme.

The planning process of any biosolid disposal alternative should start with data collection and assessment aiming to properly characterise the biosolids and the wastewater treatment system. This initial information should be compared to the applicable legal regulations for a preliminary evaluation of the feasibility of the intended application.

The following stages will involve public acceptance surveys and studies of quality and availability of future application site, means of transportation, climatic conditions, among others. Figure 8.4 shows schematically the main planning phases.

8.6.2 Preliminary planning

Preliminary planning performs an appraisal of legal and technical feasibility issues related to specific biosolid land application and systemises all pertinent information that will be needed to make the intended programme operational.

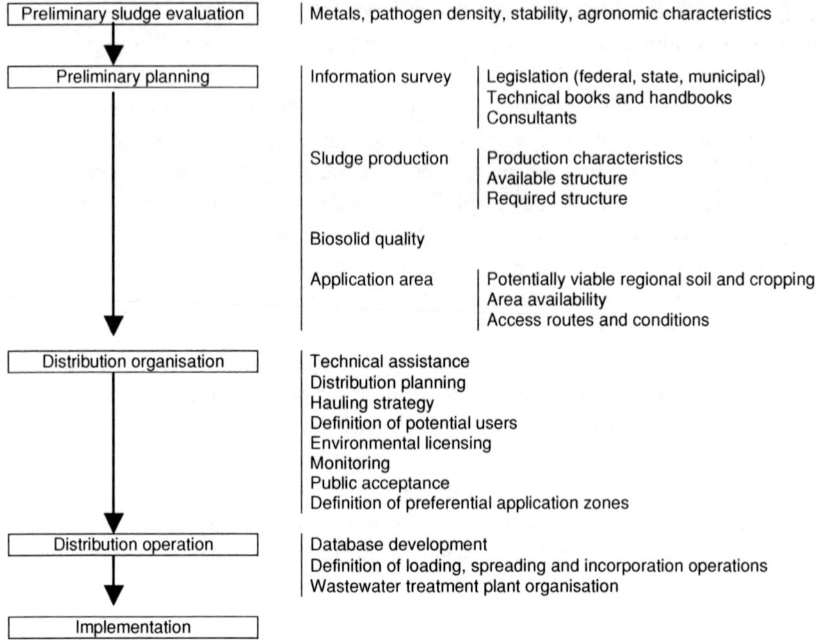

Figure 8.4. Planning of a feasibility study for land application of biosolids

8.6.2.1 Preliminary survey

All current publications covering local legislation and technical papers must be gathered and compiled with the help of skilled practitioners. The information gathered in this phase is fundamental for project development, mainly regarding legal restrictions.

8.6.2.2 Biosolids production

A report should be produced, describing in detail the wastewater treatment plants involved, covering the following items:

- *description of the treatment process* (from the inlet to the treatment plant to the sludge dewatering and final disposal, including all types of produced solids)
- *working regime* (continuous production, mixed or in batches)
- *system capacity* (design capacity, present incoming flow and potential increases)
- *available sludge handling facilities* (dewatering system, pathogen-removal system, transport/loading vehicles, laboratory, storage area, ancillary equipment, chemicals involved etc.)

- *biosolids production characteristics* (current and design volumes and physical-chemical and microbiological quality, aiming future forecasts compatible with planned sewerage system expansions)
- *available area* (sizing of existent and required area for present and future facilities, including pathogen-removal system and storage yard)
- *required area for biosolid land application and suitability of regional crops* (preliminary estimate may use Equation 8.2)

$$\text{Required area (ha)} = \frac{\text{Biosolids production (tonnes dry solids)}}{\text{Average application rate (tonnes dry solids/ha per year)}} \quad (8.2)$$

Based upon the above information, long-term planning involving material and financial resources and required area arrangements can be done and implementation guidelines may be issued.

Example 8.6

From Example 2.1 (Chapter 2), estimate the agricultural area needed for land application of dewatered biosolids from the UASB reactors. Assume an average application rate of 6 tonnes of sludge dry matter per hectare (not considering potential environmental limitations, such as metal accumulation, N leaching, pathogens dissemination).

Data: sludge production calculated in Example 2.1: 1,500 kg of SS (dry matter)/day, for a 100,000-inhabitant population served by the system.

Solution

Considering an average 6-tonne application (=6,000 kg) of dry matter per hectare, and using Equation 8.2), the following area is needed:

$$\text{Required area} = (1{,}500 \text{ tonnes/d})/(6{,}000 \text{ tonnes/ha}) = 0.25 \text{ hectares/day}$$

Example 8.7

The biosolid discharge from Example 8.6 (UASB-reactor dewatered sludge) occurs in monthly batches. Calculate the monthly area necessary for agricultural land application of the sludge.

Solution

$$\begin{aligned}\text{Monthly biosolids production} &= (1{,}500 \text{ kgSS/d}) \times (30 \text{ d/month}) \\ &= 45{,}000 \text{ kgSS/month}\end{aligned}$$

$$\text{Required area} = (45{,}000 \text{ kgSS/month})/(6{,}000 \text{ kg/ha}) = 7.5 \text{ hectares/month}$$

Example 8.8

Estimate the area required for the land application of the sludge produced per inhabitant served by the sewage treatment processes listed in Table 2.1 (Chapter 2). Remarks:

- Within the ranges presented in the table for per-capita sludge-mass daily production (kgSS/inhabitant·d), adopt the average values
- Assume application rate of 6 ton SS/ha (not considering potential environmental limitations, such as metal accumulation, N leaching, pathogens dissemination)

Solution

Using Equation 8.2 and the average per-capita SS production data from Table 2.1, the following table and graph may be produced.

Daily sludge production and required area for biosolids recycling, for various wastewater treatment systems

Systems	Sludge production (gSS/inhabitant·d)	Yearly sludge mass produced by 1,000 inhabitants (tonnes SS/year)	Agricultural area needed for disposal (ha/1,000 inhabitants·year)
Primary treatment (conventional)	40	14.6	2.4
Primary treatment (septic tanks)	25	9.1	1.5
Facultative pond	23	8.4	1.4
Anaerobic pond – facultative pond	41	15.0	2.5
Facultative aerated lagoon	11	4.0	0.7
Complete-mix aerated lagoon – sediment pond	12	4.4	0.7
Conventional activated sludge	70	25.6	4.3
Activated sludge – extended aeration	43	15.7	2.6
High-rate trickling filter	65	23.7	4.0
Submerged aerated biofilter	70	25.6	4.3
UASB reactor	15	5.5	0.9
UASB + aerobic post-treatment	26	9.5	1.6

Note: biosolid application rate = 6 tonnes SS/ha

Example 8.8 (Continued)

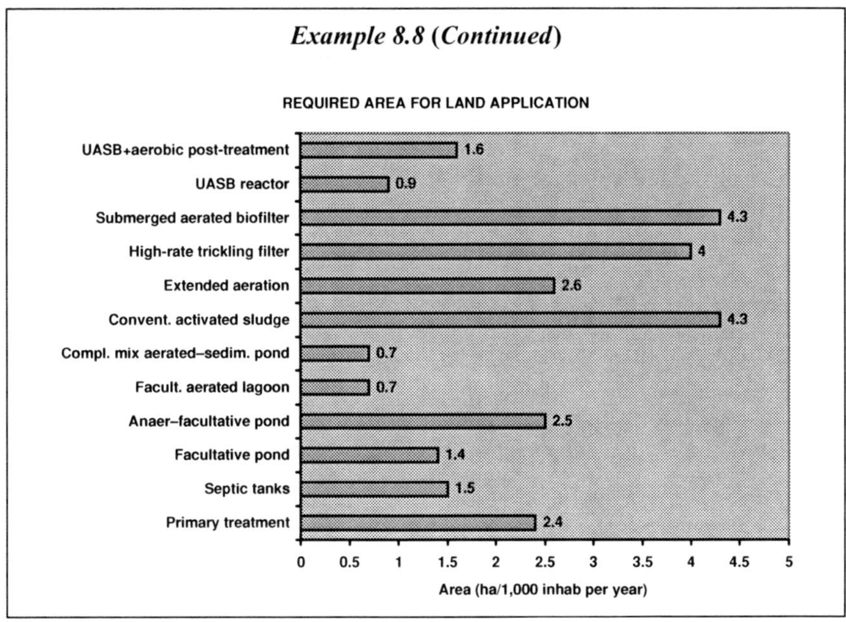

8.6.2.3 Biosolid quality

The continuous evaluation of the physical–chemical and microbiological characteristics of the biosolid, encompassing pathogen dissemination risks and soil metals accumulation, is indispensable to assess the feasibility of land application. The parameters to be evaluated must be those whose control is specified in the particular local legislation. Indispensable parameters are:

- *agronomic parameters*: N, P, K, Ca, Mg, S, C/N, pH, C
- *metals*: Cd, Cr, Cu, Zn, Pb, Ni, Hg
- *pathogen density*: viable helminths eggs, faecal (thermotolerant) coliforms
- *stability*: ash content

8.6.2.4 Application area

Application area assessment implies using all information gathered previously, including the biosolid production data, as well as regional land uses and soil characteristics, environmental restrictions and socio-economic local context.

(a) Source of information

Relevant data on land uses, regional crops and management practices can be obtained from agricultural state and municipal departments. Several such data are available on the Internet.

It is indispensable to contact the local environmental agency, for information gathering on regional environmental issues and ecological zoning. The identification of existing public water supply sources and protected watersheds is also essential for a sound biosolid management planning.

(b) Use of the soil and availability for biosolids use

Present and future land uses are always a significant issue for proper selection and definition of application areas.

Current use

The evaluation of the current use basically follows the guidelines mentioned earlier in this chapter. The municipal departments of agriculture must keep records of all relevant information regarding agriculture, reforestation, mining and other potential disposal sites, and may provide the interested party with pertinent maps, aerial photographs and satellite images of the particular region under study.

- ***Farming.*** Some farming practices such as small properties in a non-agricultural neighbourhood or traditional vegetable-growing communities may render unfeasible an application biosolids programme. Convenient places cultivate a wide variety of crops, from cereals to pasture and fruit growing, which facilitates the scheduling of the biosolid land application all year round.
- ***Reforestation.*** The cultivated forests can use high amounts of biosolids in a single application, since they are kept for long periods, thus representing a very important potential market.
- ***Reclamation areas.*** Degraded lands are found everywhere and some may be beneficially reclaimed with wastewater biosolids. The application load must consider future uses of the area, for instance, disturbed areas such as strip-mine sites are generally designed for a one-time application of biosolids based on metal loading limits. A single large application in such cases can provide organic matter and nutrients required to support establishment of a mixture of plants.
- ***Landscape gardening.*** Biosolids land application in public areas is a usual practice in developed countries and still infrequent in many developing countries. When contact with population is expected, biosolids pathogen-removal and stability processes must be stricter and demand careful observation. Private gardens can also benefit from biosolids application once careful criteria are followed.
- ***Substrate.*** Biosolids may replace the organic matter usually applied for substrate make up in soil cultivation, cuttings and flowers or organomineral fertilisers.

Future use

Future biosolid land application planning must consider preferential application zones, mainly if application sites are near densely populated areas. Those sites

Table 8.14. Climate impacts on biosolid application

Impact/Climate	Warm/Dry	Warm/Humid	Cold/Humid
Operation timing	All year round	Seasonal	Seasonal
Storage requirement	Low	High	High
Soil salinisation risk	High	Low	Moderate
Leaching potential	Low	High	Moderate
Erosion run-off risk	Low	High	High

Source: Adapted from EPA (1993)

demand more frequent monitoring, especially to avoid aesthetic problems and diminish direct or indirect contamination risks for the population. Information on those matters may have a long-term influence upon the feasibility of future application programmes.

(c) Access

Areas far away from the wastewater treatment plant or with poor access roads conditions are a major influence in the definition of preferential spreading zones.

(d) Land aptitude

Soil aptitude may be evaluated with the methodology proposed in the application example, as previously described in this chapter (see *Soil Aptitude*).

(e) Climatic features

Climatic features, also covered previously in *Soil Aptitude* in this chapter, concern application timing, soil salinization and leaching, and biosolids natural dewatering difficulties. Table 8.14 presents potential impacts of climatic changes when biosolids are land applied.

(f) Socio-economic context

In this phase definitions are required on:

- what is the regional agricultural profile and what problems may arise regarding public acceptance of the new technology
- whether regional farming equipment will be compatible with the planned work (that is, tractors, solid or liquid waste spreading devices, trucks, type of hauling practised, etc.)

8.6.3 Distribution planning

Once all pertinent data on local biosolids characteristics and volume are gathered and after knowing the pertinent agricultural regional profile, biosolid distribution may be conveniently planned and organised. Should any technical unfeasibility be detected without a perspective of acceptable solution, including legal and environmental constraints, a different wastewater sludge disposal alternative should be considered at this time.

(a) Biosolid disposal alternatives

Preliminary survey and studies on wastewater sludge disposal alternatives are expected to point out which alternatives deserve further planning. Only those complying with technical, economical, environmental and legal constraints deserve deeper consideration analysis.

Sludge production and dewatering are in-plant operations. New wastewater plant designs should consider sludge disposal alternatives, since the design of the sludge handling facilities should be consistent with the required biosolid quality, mainly, if agricultural use is involved. This is particularly true regarding the selection of the pathogen-removal process, which shall be dependent on the desired final quality and influence the entire sludge handling operation:

- better quality biosolids may eventually be used in any crop, if complying with environmental control rules
- biosolids with higher pathogen densities have crop suitability limitations, especially those that are eaten raw or whose edible parts have direct contact with the biosolid fertilised soil

According to USEPA, biosolid pathogen-removal processes may be classified into Processes to Further Reduce Pathogens (PFRP) or Processes to Significantly Reduce Pathogens (PSRP). PFRP biosolids have excellent quality and do not have any crop restriction, whereas PSRP quality imply pathogen reduction down to adequate levels, and the biosolid should comply with stricter criteria. Pathogen removal should also influence biosolid agronomic characteristics, as mentioned earlier in this chapter.

Equipment acquisition, storage and maturation area or facilities, hauling and application techniques are all dependent on the definition of the pathogen-removal system.

(b) Treatment plant structure

Once it is established how the sludge will be treated and handled, the next step is either to organise the treatment plant to accomplish the task, or to retrofit the existent facilities. Aspects that need to be taken into account are:

- equipment and labour requirements
- storage requirements
- cleaning of hauling vehicle
- supervision at the treatment plant
- transportation

(c) Public acceptance

The beneficial use of biosolids, especially its agriculture recycling, is a worldwide practice, certainly representing a very good alternative for wastewater sludge disposal, provided that due care is taken. However, it will only be feasible if supported by public acceptance, which is directly dependent upon public credibility on the safe use of the product.

The community participation must be considered as important as any technical design consideration, and the delay in the public participation may crystallise negative concepts that will hardly be overcome. Public involvement significantly diminishes the opposition to the programme. The goals of a public participation programme are:

- To increase awareness of technicians, scientists, users and consumers on advantages and precautions to be observed when beneficial uses of biosolids are considered
- To increase consciousness of the affected population about all process stages, stressing measures taken to assure that biosolid quality will not jeopardise public health
- To ask technicians, community leaders and politicians opinions and suggestions
- To assure public access on biosolid quality control results and impacts regarding areas fertilised with biosolids

Public education and interactivity are a major help to attain such goals. The educational programme must impartially approach advantages and disadvantages of biosolids use. Topics to be presented are:

- reasons for recycling option among other sludge disposal alternatives, such as incineration and landfills
- measures taken assuring that biosolid production, handling and application are safe operations
- crop and soil restrictions
- costs involved
- project advantages considering the economical benefits to farmers
- project advantages considering environmental improvement
- comprehensive description of the whole process, from biosolid generation up to its land application

The key to a successful interactivity lies on a direct and open communication channel between the involved community and the project technicians.

Marketing strategies should consider how to overcome the absence of correct knowledge on benefits and potential risks regarding biosolids land application. Prejudice against land application of the sludge is a common position, which makes marketing very important since the beginning of the project.

(d) Technical assistance

Agronomic assistance is an essential tool for a successful biosolid land application programme. The agronomic engineer or a specialised contractor should be responsible for the selection of the properties in which the biosolids will be applied and for technical advising on biosolids use. It is important to keep in mind the primary responsibility of the sanitation company regarding any problems that may

Table 8.15. Alternatives for agronomic assistance

Strategy	Company	Advantages	Disadvantages
Association	State agency providing agricultural technical assistance	Credibility with the producer Multi- disciplinary technical support Local offices in almost every municipality Training ease Experience in technology spreading Regional knowledge Willingness to work with small producers	Bureaucracy Difficulty to achieve a profitable relationship with large producers
Business transactions	Private company providing agriculture technical assistance	Training easiness Contact with the local rural area Experience and interest in technology spreading Easiness to achieve a profitable relationship with large growers	Starting of a novel activity Restrict technical staff Lack of homogeneity from technicians from different municipalities Cost
	Specifically assigned professional to do the work	The own company rules the technical assistance Consistent selection criteria	Cost
Records	Autonomous professionals	Payment per area, number of prescriptions or biosolids applied volume	Lack of homogeneity in staff Lack of standardisation in information and reports

eventually happen regarding this alternative of sludge disposal. Table 8.15 shows strategic alternatives, advantages and disadvantages of technical agronomic assistance programmes to farmers.

(e) Monitoring

Monitoring is an indispensable tool to assess whether positive or negative impacts are occurring from the current biosolids land application process. It helps maximise positive impacts and propose control measures for the negative ones. The monitoring programme must generate sufficient data to make biosolid recycling an environmentally and socially suitable operation in full compliance with legal parameters. Monitoring is covered in more detail in Chapter 10.

(f) Environmental licensing

The environmental operational license to start-up the wastewater treatment plant must be given only when the wastewater sludge disposal option has been approved by the local environmental agency. Usually, the sanitation company is required to file a complete set of documents encompassing the biosolids disposal plan: sludge production flowsheet, dewatering, pathogen removal and handling, plus quality control and monitoring routines.

8.6.4 Distribution operation

The last planning phase covers the arrangements for implementing the activity involving staff selection and training, control programme adjustments, report and data record systems:

- **Selection and training of technical staff.** All operational and managerial personnel must be hired and trained in full compliance with the legislation and the environmental agencies requirements.
- **Control programme.** The control programme is a map of the recycling activity, where each programme stage is clearly represented, including risk maps with detailed contingency measures to be taken should an emergency happen.
- **Report and data record system.** Reports on the biosolid land application must be submitted to environmental agencies, showing monitoring results. The area utilisation history is built up and may be available for public consultation, amplifying reliance on competence of the responsible staff in charge of the sludge disposal. Those reports are usually filed for some years and encompass sludge treatment, pathogen removal, biosolids handling and land application. The producer is always responsible for the environmental effects of his waste. All areas where sludge has been applied should be registered, with control numbers for the particular biosolid lots handled, as well as their characterisation. Table 8.16 shows information to be registered and reported.

8.7 LANDFARMING

8.7.1 Preliminaries

In the sludge land disposal and treatment system known as *landfarming*, there is no productive use of sludge nutrients and organic matter. The process goal is the sludge biodegradation by soil microorganisms present in the tillable profile, while metals are held on top soil surface layers.

Soil supports microorganisms and oxidation reactions of the organic matter. As the area dedicated to landfarming does not aim at any crop cultivation, rates applied are much higher than those with agronomic purposes. Nevertheless, a number of environmental concerns are valid for both landfarming and agriculture, although with different limits since landfarming is associated with greater technological interventions to control environmental pollution.

Technically, it is feasible to promote compaction and impermeability of the soil layer down to a depth of 60–70 cm from the surface, to build a proper drainage

Table 8.16. Data recording and reporting

Activity	Information	Technician responsible: Production	Technician responsible: Assistance	Information that must be reported
Biosolid production	Environmental permits for biosolid land application	√		√
	Distribution plan	√		
	Metal content	√		√
	Pathogen density	√		√
	Biological stability	√		√
	Description of stabilisation and pathogen removal	√		
	Biosolids quantity allowed to leave treatment plant	√		√
	Hauling records	√		
	Responsibility Statement (signed by user)	√		
Land application	Application site selection		√	√
	Area size		√	√
	Application date		√	√
	Amount of applied sludge		√	√
	Area description and aptitude records		√	√
	Handling practices and description records		√	√
	Soil pollutant accumulation records		√	√

system, and to collect and treat all percolates. Sludge application rates may then increase because nitrates subsoil contamination is being prevented. If these techniques are not applied, a greater groundwater quality control is necessary and sludge application rates would be lesser.

As the land farming soil is continuously revolved to provide aeration, vegetation usually does not grow in such sites, although some landfarming areas have species with the sole purpose to fix nutrients and increase evapotranspiration.

8.7.2 Basic concepts

Since the process objective is organic waste biodegradation on top soil, the pertinent parameters to define process efficiency are temperature, rainfall, soil pH, aeration, nutrient balance, soil physical state and sludge characteristics.

Due to the high sludge application rates lasting for several years, the main environmental concerns are related with the possible contamination of waters (both underground and surface). Therefore, an efficient and well-maintained surface drainage system should be provided from the beginning of the operation. Other negative environmental impacts like odours and undesirable attraction of vectors may occur. Figure 8.5 shows a cross-section of a landfarming cell.

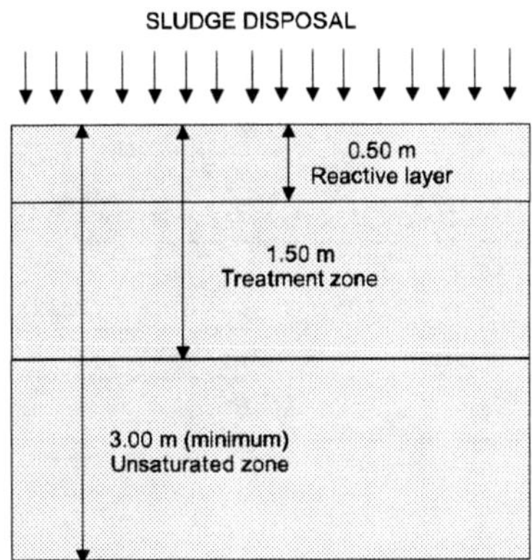

Figure 8.5. Schematic cross-section of a landfarming cell

In the reactive layer, the sludge bio-oxidation reactions occur through intense microbial activity during the biodegradation process. Periodically, it must be tilled to promote aeration and then harrowed levelled.

The 'treatment zone' is unsaturated by definition and should have no more than 1.50 m depth. This is where released components from biodegradation are fixed and transformed.

Water table should be at least 3.0 m deep during the wet weather season. Water table depths of 10–20 m are not unusual and are the ideal situation for process implementation. The deeper the water table, the safer will be the area.

The sludge application rate will be a function of the soil biodegradation capacity, which, as mentioned, depends on several factors. Organic matter biodegradation rate may be estimated by Bartha's respirometric test. If organic matter biodegradation rate is known, waste treatability can be evaluated and best handling conditions may be selected, such as application rate, soil pH correction, ideal soil moisture and nutrient balancing.

The wastewater sludge to be tested shall be previously characterised with acute toxicity tests (if required by the environmental agency), fixed and volatile total solids, moisture content, organic carbon, nitrogen, phosphorus, heavy metals, oil and greases.

8.7.3 Selection of areas for landfarming

Most criteria given to select suitable biosolids land application sites for agricultural purposes are also pertinent to choose landfarming sites, with some particular considerations. According to EPA 625/1-83-016 (EPA, 1983), the following aspects must be considered during site selection for landfarming:

- distance of the area from the sludge production site
- condition of the transport roads

Table 8.17. Landfarming site selection criteria: main USEPA parameters

Parameter	Unacceptable	Desired
Slope	>12%	<3%
Soil permeability	$>1 \times 10^{-5}$ cm/s for soil layers less than 0.6m deep	$\leq 1 \times 10^{-7}$ cm/s for soil layers deeper than 3.0 m
Distance from surface waters	<90 m	>300 m >60 m for intermittent streams
Water table depth	<3 m	>15 m
Distance from drinking wells	<300 m	>600 m

- existence of an impermeable geological barrier, rocky-layer type, avoiding fracture zones
- absence of nearby aquifers
- transition distance between the selected site and populated areas, public interest sites or wells
- distance from water courses
- suitability of the topography
- climate conditions favouring high evaporation and transpiration rates
- well-drained soils, high cation exchange capacity and pH above 6.5

The selected area should be free from 100-year return period floods (ABNT, 1997) and should comply with soil regional uses as prescribed by law.

Table 8.17 summarises some parameters specified by EPA–625/1-083-016.

Regarding soil permeability, it should be high in the reactive layer to avoid puddles and anaerobic conditions. On the other hand, deeper than 3.0 m, low permeability (less than 1×10^{-7}cm/s) is desired to prevent infiltration. Such a water barrier can exist naturally or be made up by soil compaction or a synthetic liner.

8.7.4 Design and operational aspects

The required area must be determined in accordance with the respirometric test results and should be subdivided into several cells for easier environmental monitoring and control of the application and resting periods.

As soil conditions must be kept aerobic, if the spread sludge has a high moisture content, the application rates must be smaller.

Landfarming designs should specify sludge application rates, area handling, sludge application technology, sludge application frequency, equipment provided for process operation and soil handling regarding fertilisation and pH correction. They should also include storm water drainage system details and percolate system collection, including its corresponding treatment, if required.

A buffer area should also be sized for contingencial sludge storage, according to the particular site operational characteristics. Similar to sanitary landfills, landfarming area projects must define appurtenances such as fences, balances, inner access roads, field office, shed and workshop for machine maintenance compatible with the site dimensions. The applied sludge should be surface spread and incorporated by an agricultural harrow.

The project must also include an environmental monitoring plan, an emergency plan and a closure plan.

8.7.5 Environmental monitoring

The unsaturated and also the saturated zones should be monitored. Both the soil and the soil solution must be monitored in the unsaturated zone to verify whether some migration from the treatment zone is occurring. The parameters subjected to monitoring should be selected in accordance with the sludge characteristics and must comply with the environmental agency requirements.

Information on the soil solution quality and the chemical soil composition below the treatment zone should be obtained from the sampling and analysis undertaken. All results shall comply with the environmental agency requirements.

The landfarming area must be built and operated preserving the groundwater quality. Groundwater monitoring should be carried out unless exempted by the environmental agency. There should be a sufficient number of installed monitoring wells, so samples withdrawn are able to truly represent the aquifer water quality. Brazilian practice (ABNT, 1997) recommends:

- The monitoring well system must consist of at least four wells: one upstream and three downstream following the direction of the preferential draining pathway of the groundwater.
- The wells must have a minimum diameter of 10 cm (4 inch) for proper sampling. To avoid contamination they must be lined and top covered.

The monitoring programme should:

- indicate the parameters to be monitored, considering the sludge characteristics and mobility of the components
- establish procedures for sampling and preservation of samples
- establish background values for all parameters of the programme. These values may come from upstream sampling before the beginning of the operation and may be later on compared to water quality results after regular start-up. Background tests have the advantage of determining when a particular contamination has occurred and whether it may be attributed or not to the start-up of a particular facility

8.7.6 Closure plan

After the area closure, the continuity of the operation in the treatment zone must be assured to achieve the highest possible decomposition, transformation and fixation of the applied constituents. The draining systems of rainwater and percolates must also be kept operational, as well as the effluent treatment system, if existent. After closure, agricultural use of the area must not be allowed. Monitoring of the unsaturated zone must continue for one year after the last application.

Example 8.9

From Example 2.1 (Chapter 2), estimate the area needed for the sludge treatment by landfarming. The sludge comes from a treatment plant using anaerobic sludge blanket reactor (UASB) for a population of 100,000 inhabitants. Assume an application rate of 300 tonnes of sludge/ha·year (dry basis), based on Bartha's respirometric test.

Data: sludge production from Example 2.1: 1,500 kg of SS/d.

Solution:

Yearly sludge production: 1,500 kgSS/d × 365 d/year = 547,500 kgSS/year
Application rate: 300,000 kgSS/ha year
Required area: 547,500/300,000 = 1.825 ha or 18,250 m^2

Although the landfarming area is much smaller than the agricultural recycling area (Example 8.6), it is important to notice that a number of technical and control appurtenances are required for landfarming, such as drainage systems, banks, edgings, subsurface waterproofing, monitoring wells, etc. These are not needed for biosolid land application in agricultural areas.

9

Sludge transformation and disposal methods

M. Luduvice, F. Fernandes

9.1 INTRODUCTION

Worldwide urbanisation leads to the growth of large metropolitan areas, imposing constraints, amongst others, for sludge disposal alternatives. Freight costs and the adverse heavy traffic impact in metropolitan areas favours the adoption of sludge treatment and disposal alternatives within the wastewater treatment plant area. This scenario justifies the consideration of processes such as incineration and wet air oxidation.

Regardless of the adopted technology, all treatment and disposal processes present advantages and disadvantages, many of them regarding possible contamination of soil, receiving water bodies and atmosphere. Sludge combustion causes serious concerns towards atmospheric pollution and safe disposal of residual ashes.

Many countries are increasingly using the incineration process as a response to growing difficulties in maintaining landfills as a final sludge disposal route due to the increasing competition for space in landfills, disposal costs, legislation constraints and incentives to sludge recycling.

ISBN: 1 84339 166 X. Published by IWA Publishing, London, UK.

The present chapter deals with the following sludge transformation and disposal methods:

- thermal drying
- wet air oxidation
- incineration
- landfill disposal

Some of these processes are also described in other chapters of this book (Chapters 4 and 6). From the above alternatives, only landfills can be classified as a final disposal route, since the others, although presenting a high water-removal capability, still leave residues that require final disposal.

Landfills should only be selected as a final disposal alternative if wastewater sludge use is an unfeasible solution. Most municipal wastewater treatment plant sludges have physical–chemical properties useful for agriculture or industrial use, allowing the development of a more constructive attitude towards sludge. Under certain circumstances, sludge may be looked upon as a commodity and not as a useless residue.

9.2 THERMAL DRYING

Thermal drying is a highly flexible process, easily adapted to produce pellets for agricultural reuse, sanitary landfills disposal or incineration. The process applies heat to evaporate sludge moisture. The produced pellets can be used as fuel for boilers, industrial heaters, cement kilns and others. Pellet solids concentration varies from 65–95% (5%–35% water content). Main advantages of thermal sludge drying are:

- significant reduction in sludge volume
- reduction in storage and freight costs
- stabilised final product easily transported, stored and handled
- final product free of pathogenic organisms
- final product preserving the characteristics of soil amendment from sewage sludge
- final product suitable for unrestricted agronomic reuse, incineration or final disposal in landfills
- possibility of accommodation in small size packages

Main limitations of thermal drying processes are:

- production of liquid effluents
- release of gases into the atmosphere
- risk of foul odour and disturbing noise

Thermal drying processes may be classified as *indirect*, *direct* or *mixed*. Indirect processes produce pellets with up to 85% solids concentration. For solids contents higher than 90% and possible production of organomineral fertilisers, direct drying processes are recommended.

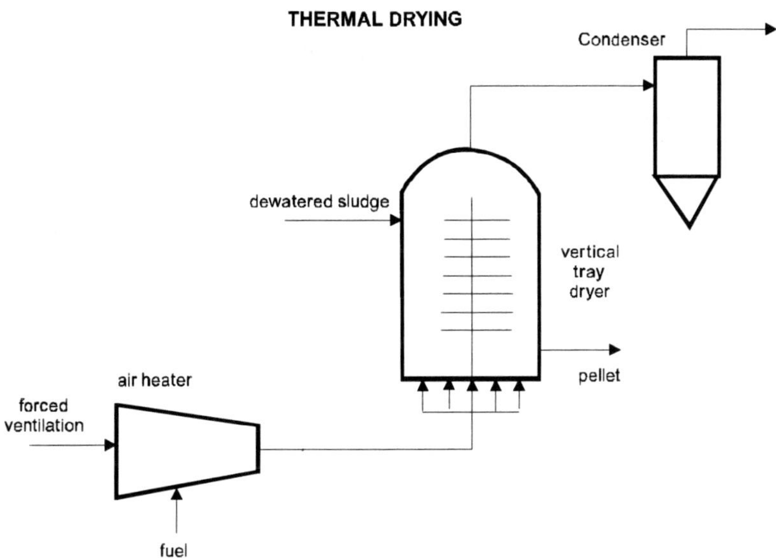

Figure 9.1. Thermal drying process operation

Liquid effluent is less than 1% of the total treatment plant flow and may be recycled to the plant headworks, provided sufficient capacity is available to deal with the additional organic load. When thermal drying anaerobic-sludge, surplus ammonia nitrogen may become a problem during liquid effluent treatment.

Both direct and indirect drying processes produce gaseous emissions with foul odours potential. The odour associated with the dry product, although less intense, is similar to the original sludge odour. It is highly recommended that the drying unit be isolated, preferably under a negative pressure environment to minimise gaseous release hazards.

Figure 9.1 presents a thermal drying process operation.

9.3 WET AIR OXIDATION

Originally developed in Norway for paper industry residues treatment, wet air oxidation was adapted for sewage sludge treatment in the United States during the 1960s. Despite its promising start, it did not achieve the expected results and was later adapted to treat high-toxicity industrial liquid wastewater. Wet oxidation is recommended when the effluent is too diluted to be incinerated, and toxic/refractory to be submitted to biological treatment.

The process is based on the capability of dissolved or particulate organic matter present in a liquid to be oxidised at temperatures in the range of 100 °C–374 °C (water critical point). The temperature of 374 °C limits the water existence in liquid form, even at high pressures. Oxidation is accelerated by the high solubility of oxygen in aqueous solutions at high temperatures. The process is highly efficient in organic matter destruction of effluents in the 1%–20% solids concentration range, allowing enough organic matter to increase the reactor internal temperature through heat generation without external energy supply. The upper 200 g/L (20%) solids concentration limit avoids the surplus heat to raise the temperature above the critical value, which could lead to complete evaporation of the liquid. Wet air oxidation of organic matter can be described by Equation 9.1.

$$C_aH_bO_cN_dS_eCl_f + O_2 \rightarrow CO_2 + H_2O + NH_4^+ + SO_4^{2-} + Cl^- \quad (9.1)$$

Theoretically, all carbon and hydrogen present can be oxidised to carbon dioxide and water, although factors such as reactor internal temperature, detention time and effluent characteristics influence the oxidation degree achieved. As can be noticed from Equation 9.1, organic nitrogen is converted into ammonia, sulphur into sulphate, and halogenated elements into their Cl^-, Br^-, $I^-$and Fl^- ions. These ions remain dissolved, and there is no production of sulphur or nitrogen oxides (SOx and NOx).

Due to the exothermic characteristic of Equation 9.1, the wet air oxidation process is able to produce sufficient energy to maintain a self-sustaining process. The autogenous operation calls for influent COD concentrations higher than 10 g/L, only a fraction of the 400 g COD /L concentration required to maintain autogenous operation in incinerators.

Recent technological developments and restrictions imposed by environmental legislation to final sludge disposal in several countries have renewed the interest in wet air oxidation use for sewage sludge stabilisation. Wet air oxidation processes are currently, once again, being considered for wastewater treatment plants serving large metropolitan areas.

Sewage sludge organic matter, when submitted to wet air oxidation, may be considered *easily oxidisable* or *not easily oxidisable*. In the first category are proteins, lipids, sugars and fibres, the most usual constituents of sludge, which account for approximately 60% of the total organic matter.

The mains control variables of the wet air oxidation process are (a) temperature, (b) pressure, (c) air/oxygen supply and (d) solids concentration.

The process may be classified according to the working pressure as:

- *low pressure oxidation*
- *intermediate pressure oxidation*
- *high pressure oxidation*

The main purpose of low-pressure wet air oxidation is to reduce sludge volume and increase its dewaterability for thermal treatment, whereas intermediate and high-pressure oxidation are conceived to reduce sludge volume through oxidation of volatile organic matter into CO_2 and water.

In spite of its efficiency, wet air oxidation process is far from being a complete process and its application at an industrial scale requires efficient operation and maintenance. The most usual problems in industrial scale are:

- foul odours
- corrosion of heat exchangers and reactors

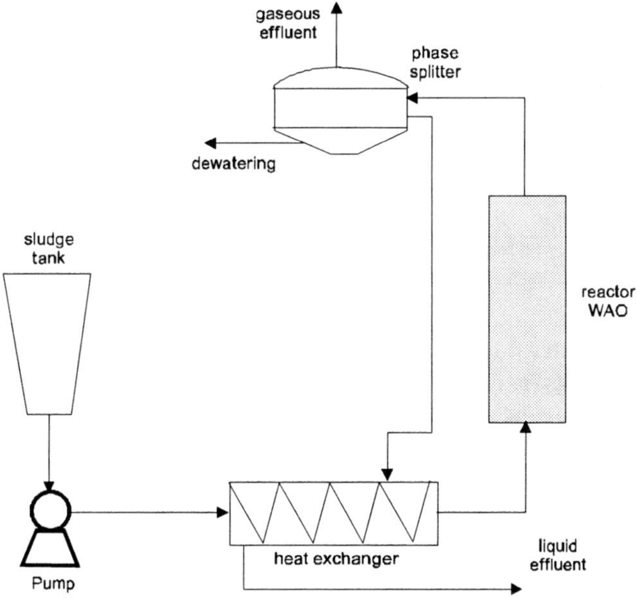

Figure 9.2. Conventional wet air oxidation system with a vertical reactor

- required power consumption to start-up the oxidation process
- high COD in liquid effluent
- high metal content in residual ashes

Figure 9.2 shows a vertical reactor wet air oxidation system. The influent sludge is pumped towards the Wet Air Oxidation (WAO) reactor, passing through a heat exchanger to raise its temperature. The WAO reactor effluent goes through a phase splitter, routing the sludge for dewatering, whereas the liquid flows back through the heat exchanger, where part of the heat is transmitted to the incoming sludge. The gaseous effluent is released into the atmosphere after being treated by an electrostatic precipitator and filtered for solid particles and odorous substances removal.

Wet air oxidation may use air or pure oxygen as oxygen supply. Compressed air as an oxidising agent is usually found in wastewater treatment plants.

Comparative studies for effluent treatment with up to 20% of solids have shown that capital costs for a wet air oxidation process are higher than those required for an incinerator, although the operational cost of the former is significantly lower, due to less external energy required. Wet air oxidation can treat almost any type of organic sludge produced at domestic or industrial wastewater treatment plants. The solid produced is sterile, not putrescible, settles readily and may be easily mechanically dewatered. Solids low nitrogen concentration and fairly high metals

Table 9.1. Typical operational ranges of wet air oxidation treatment of sewage sludges

Parameter	Type of oxidation process: Low pressure (thermal treatment)	Intermediate pressure	High pressure
Pressure (atm)	20.5–27.3	27.3–54.6	54.6–136
Temperature (°C)	148–204	204–260	260–315
Organic matter destruction (%)	5–10	10–50	50–90
Volume reduction (%)	25–35	30–60	60–80
Sludge sterilisation	yes	yes	yes
Autothermal reaction	no	yes	yes
Improvement in dewaterability	yes	yes	yes

content may render it unsuitable for land application aiming agricultural reuse, adding constraints for its final destination.

The gaseous output from a wet air oxidation process is a mixture of nitrogen, oxygen, carbon dioxide and hydrocarbons. The release of foul odours is directly dependent on the oxidation degree achieved inside the reactor.

A significant COD removal occurs and a large portion of it is transformed into low-molecular-weight volatile acids (e.g., acetic acid, propionic acid and others), which may reach COD values around 5,000–10,000 mg/L.

The liquid phase from intermediate and high-pressure wet air oxidation has smaller organic content and may be biologically treated. Its organic content is composed of low-molecular-weight volatile acids and amino acids. The COD:BOD ratio is about 2 and removal efficiencies higher than 80%–95% for COD and BOD are achievable. The process supernatant may be biologically treated, either by anaerobic reactors or activated sludge.

Table 9.1 presents a comparison among the main wet air oxidation processes used for sewage sludge treatment.

Wet air oxidation units are considered highly sophisticated, requiring skilled personnel for operation and maintenance. All reactor material should be made of stainless steel 316 to avoid corrosion from formed acids.

Wet air oxidation technology has been combined with the activated sludge process (Deep Shaft Technology), with reactors going down as deep as 1.6 km to achieve sludge treatment. Wet air oxidation in deep shafts makes pumps, heat exchangers and high-pressure reactors unnecessary items, which significantly diminishes capital costs. Deep Shaft Technology has been recommended where space is at a premium.

9.4 INCINERATION

Incineration is the sludge stabilisation process which provides the greatest volume reduction. The residual ashes volume is usually less than 4% of the dewatered sludge volume fed to incineration. Incinerators may receive sludge from several treatment plants and are usually built to attend over 500,000 population equivalents, with capacities higher than 1 tonne/hour.

Table 9.2. Calorific power of different types of sewage sludge

Type of sludge	Calorific power (kJ/kg dry solids)
Raw primary sludge	23,300–29,000
Anaerobically digested sludge	12,793
Activated sludge	19,770–23,300

Source: Adapted from WEF (1996)

Sludge incineration destroys organic substances and pathogenic organisms through combustion obtained in the presence of excess oxygen. Incinerators must use sophisticated filter systems to significantly reduce pollutant emissions. Gases released to the atmosphere should be regularly monitored to ensure operational efficiency and safety.

Incinerator design requires detailed mass and energy balances. In spite of the considerable concentration of organic matter found in dewatered sludge, sludge combustion is only autogenous when solids concentration is higher than 35%. Dewatered cakes with 20 to 30% total solids can be burned with auxiliary fuels, such as boiler fuel having low sulphur content. The calorific value of sludge is fundamental in reducing fuel consumption. The combustible components found in sludge are carbon, sulphur and hydrogen, present as fat, carbohydrates and proteins. Table 9.2 shows typical calorific values of different types of sludge.

Products from complete combustion of sludge are water vapour, carbon dioxide, sulphur dioxide and inert ashes. Good combustion requires an adequate fuel/oxygen mixture. Oxygen requirement for complete combustion is usually much higher than the stoichiometric value of 4.6 kg of air for every kg of O_2. Generally, 35%–100% more air is required to assure complete combustion, the necessary excess air depending upon the sludge characteristics and the type of incinerator.

The amount of oxygen needed for complete combustion of the organic matter can be determined from the identification of the organic compounds and from the assumption that all carbon and hydrogen are oxidised into carbon dioxide and water. The theoretical formula can be expressed as:

$$C_aH_bO_cN_d + (a + 0.25b - 0.5d)\,O_2 \rightarrow aCO_2 + 0.5cH_2O + 0.5dN_2 \qquad (9.2)$$

Two types of incinerators are currently in use for sewage sludge:

- *multiple chamber incinerator*
- *fluidised bed incinerator*

A multiple chamber incinerator is divided into three distinct combustion zones. The higher zone, where final moisture removal occurs, the intermediate zone where combustion takes place and the lower or cooling zone. Should supplementary fuel be required, gas or fuel oil burners are installed in the intermediate chamber.

A fluidised bed incinerator consists of a single-chamber cylindrical vessel with refractory walls. The organic particles of the dewatered sludge remain in contact with the fluidised sand bed until complete combustion.

Table 9.3. Example of the influence of solids concentration in a fluidised bed incinerator operation

Parameter	Sludge with 20% solids	Sludge with 26% solids
Volatile organics (%)	75	75
Available energy (MJ/mt DS)	3,489	4,536
Exhaustion temperature (°C)	815	815
Excess air required (%)	40	40
Air temperature (°C)	537	537
Fuel consumption (L/mt DS)	184	8
Processing capacity (kg DS/hour)	998	1,361
Power (kWh/ mt DS)	284	207
Gas washing water (L/mt DS·s)	30	26
Operating time (hour/d)	18.2	13.3

Source: Adapted from WEF (1992). DS = dry solids; mt = metric ton = 1,000 kg.

The present trend favours fluidised bed incinerator over multiple chamber furnaces, due to smaller operational costs and better air quality released through its chimney. Operation under autogenous conditions at temperatures above 815 °C assures complete destruction of volatile organic compounds at cost-effective price. Dewatering equipment nowadays is able to feed cakes higher than 35% total solids to incinerators, making autogenous combustion operation feasible. Table 9.3 illustrates the advantage of feeding sludge with higher solids concentration to a fluidised bed incinerator.

Use of incinerators in sludge treatment is restricted to wastewater treatment plants serving large urban areas due to high costs and sophisticated operation involved. Nevertheless, restrictions to agriculture sludge reuse caused by excessive metals concentrations, long hauling distance and volume constraints in urban landfills may favour incineration as a viable alternative for wastewater sludge treatment.

Atmospheric emissions from incinerators are controlled by optimising the combustion process and using air filters. Air pollutants released consist of solids evaporated or compounds formed during combustion, the main ones being:

- nitrogen oxides (NOx)
- incomplete combustion products – carbon monoxide (CO), dioxins, furans, etc.
- acidic gases: sulphur dioxide, hydrochloric acid and hydrofluoric acid
- volatile organic compounds: toluene, chlorinated solvents

Solids are also present in atmospheric emissions from incinerators, consisting of thin particulate matter made up of metals and suspended solids condensable at room temperature. The metals concentration in suspended solids is directly dependent on the incinerated sludge quality. Electrostatic precipitators are widely used for removal of particulate matter from incinerator emissions.

In spite of considerable reduction in sludge volume, incineration cannot be considered a final disposal route, as residual ashes require an adequate final disposal.

Table 9.4. Typical composition of ashes from wastewater sludge incineration

Component	Composition (in dry weight)
SiO_2	55%
Al_2O_3	18.4%
P_2O_5	6.9%
Fe_2O_3	5.8%
CaO	5.4%
Cu	650 mg/kg
Zn	450 mg/kg
Ni	100 mg/kg
Cd	11 mg/kg

Efficient combustion assures complete destruction of organic matter present in the ashes inert inorganic matter with a considerable concentration of metals. The quantity of residual ashes varies according to the sludge being incinerated. For raw sludge 200–400 kg/tonne may be expected, whereas for digested sludge, 350–500 kg/tonne might be produced, due to the smaller concentration of volatile solids. Table 9.4 shows typical composition of ashes from wastewater sludge incineration.

Risks of inadequate ashes disposal are associated with the possible leaching of metals and their later absorption by plants. Final disposal in landfills is the most suitable ash disposal alternative, bearing in mind that disposal onto soil is not advisable. More recent technologies use a cement and ashes mixture, assuring reliable metals retention.

It is also possible to co-incinerate sludge in cement kilns or in thermoelectric power plants using mineral coal as fuel. Co-incineration reduces incineration capital as well as operational and maintenance costs, since they are integrated to the industrial process train.

Figure 9.3 shows a fluidised bed incinerator with washing system and gas cooling.

9.5 LANDFILL DISPOSAL

9.5.1 General considerations

Landfill is a technique for safe disposal of solid urban refuse onto soil, with no damage to public health and minimum environmental impacts, using engineering methods able to confine the disposed waste within the least possible area and smallest possible volume, covered with a soil layer after each working day, or at smaller time intervals, if necessary (ABNT, 1992).

For sludge disposal in landfills there is no concern regarding nutrients recovery or sludge use for any practical purpose. Anaerobic biodegradation takes place in sludge confined within cells, generating several by-products, including methane.

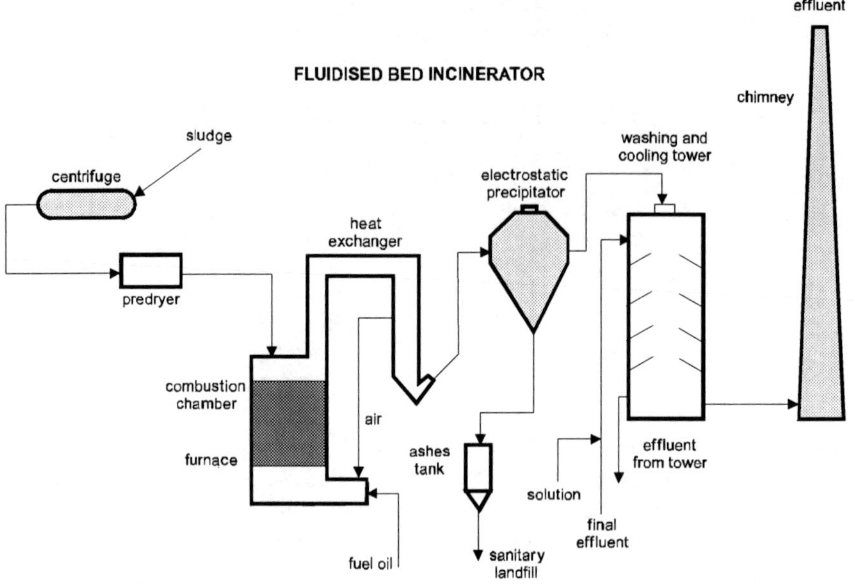

Figure 9.3. A fluidised bed incinerator with washing and cooling gas systems (adapted from CIWEM, 1999)

Sludge disposal into landfill depends on sludge properties and landfill characteristics. Two types of landfill disposal may be considered:

- **Exclusive (dedicated) sanitary landfills**: especially designed and constructed to receive sewage sludge, incorporating special features to cope with specific sludge properties and to comply with environmental constraints. Usually requires thermally dried sludges or cakes with high solids contents (>30%).
- **Co-disposal with urban solid waste**: wastewater sludge is disposed of in a landfill with municipal solid wastes. Mixing of sludge with urban wastes tends to accelerate the biodegradation process as a function of the nitrogen content and the sludge inoculation potential. The inconvenience of this alternative is the reduction of landfill lifetime if the amount of sludge is significant.

There are a significant number of technologies available to build, operate and maintain both types of landfills, as alternatives to land disposal of sludge.

Sludge from domestic wastewater treatment may be considered a non-inert residue, and is generally classified as non-hazardous residue. As a matter of fact, a number of sewage sludge samples have undergone waste extraction tests or metal leaching tests and solubilisation tests (Santos, 1996), demonstrating that sludge from municipal treatment plants is not a hazardous waste. Conversely, if wastewater contains a high concentration of industrial effluents, sludge may become heavily contaminated, requiring disposal at a landfill site licensed to handle hazardous waste.

Landfills are a flexible solution as they may accommodate variable sludge volumes, absorb excess demands from other forms of final destination and operate independently from external factors. Sludge characteristics such as degree of stabilisation or pathogen level are not of primary concern while choosing landfill as a final disposal route.

An important consideration for a monofill implementation is availability of suitable land not far from the wastewater treatment plant. Site selection should be based on an extensive study applying multi-disciplinary criteria to locate the best environmental and economical option.

Besides approval from the environment agency and full compliance with stringent standards, neighbouring population of a future landfill site should be listened to and have their concerns taken into consideration during the design and construction phases.

9.5.2 Area selection and environmental impact considerations

One of the first activities that must be accomplished in a sanitary landfill project is an extensive evaluation of the environmental impacts associated with different implementation and operational phases of the landfill. Environmental assessment allows project definition of protection measures necessary to control and minimise negative impacts. Landfill site selection is critical, and several impacts may be eliminated or minimised if the selected site presents favourable characteristics. Landfill, when not properly designed or operated, may cause pollution to:

- air, through foul odours, toxic gases or particulate material
- surface water bodies, through percolate drainage or sludge transport by run-off
- soil and groundwater, by infiltration of percolated liquids

Table 9.5 presents a list of main environmental impacts one should take into consideration while searching for a suitable area to locate a landfill site.

A landfill site selection process should take into consideration the following steps:

- selection of macro-regions, considering access and waste generating points
- identification of all legal constraints within macro-regions and exclusion of affected areas
- preliminary evaluation of the remaining macro-regions, analysing:
  - topographic requirements: dry valleys and hillsides with slopes below 20%
  - geological and hydrogeological requirements: low-permeability lithology, not excessively fractured, soils not excessively deformable, ideally located near the watershed limit, and water table depth greater than 1.50 m

Table 9.5. Main environmental aspects for selection of landfills sites

Aspect to be considered	Characteristics to be evaluated
Surface and groundwater	Site geology and hydrology Localisation of surface water bodies Site location within the watershed and local use of water resources Local climate
Air	Local climate Direction of prevailing winds Distance and transition areas to housing developments
Soil	Soils characteristics Local flora and fauna Site geology and hydrology
Anthropic environment	Landscape changes Aesthetic changes Distance from housing developments Direction of prevailing winds Change in land value Local legislation

Source: Gómez (1998)

- preliminary selection using aerial photos to determine favourable sites located in the remaining areas that comply with the selective criteria listed above
- field survey and preliminary selection taking into account all gathered information
- development of technical studies comparing all potential areas
- environmental licensing of the selected landfill site

Although site selection is always complex, the level of complexity and design detail to be considered during this phase is a function of the volume of sludge to be disposed of.

9.5.3 Exclusive landfills or monofills

Exclusive landfills, dedicated landfills or monofills are designed to exclusively receive wastewater sludge. Most exclusive landfills in the United States use trenches, with 1–15 m width (Malina, 1993). Narrow trenches (1–3 m wide) allow truck unloading without vehicle traffic onto the disposal ditch. When narrow trenches are used, sludge total solids content can be lower than 30%, since it will be supported by the trench side walls. This kind of landfill requires large areas but allows operational simplicity and is recommended for small sludge volumes.

Large trenches (3–15 m wide) allow trucks' access to the disposal ditches to unload sludge. They require solids concentrations higher than 40% to support vehicle traffic (Nogueira and Santos, 1995).

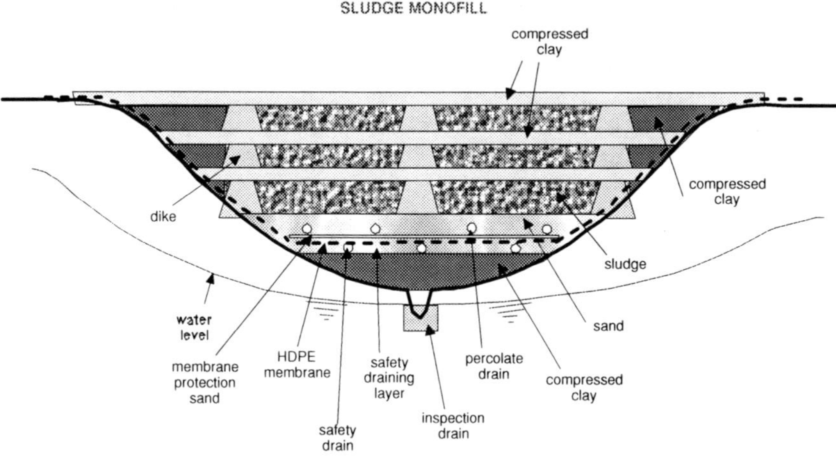

Figure 9.4. Cross section of a large sewage sludge monofill, comprising over-layered cells and dikes

Narrow trenches may accommodate 450–2,100 tonnes of sludge cakes (dry basis) per ha, including areas between trenches. Wide trenches, on the other hand, may be landfilled with 1,200–5,500 tonnes/ha (Malina, 1993).

Both concepts based on trenches are simple, proven technologies. Large size landfills require alternative engineering solution to allow the disposal of large volumes of sludge cakes in a relatively small area. Figure 9.4 presents an example of a landfill, with large over-layered cells spaced by dikes.

9.5.4 Co-disposal with municipal solid wastes

Co-disposal with municipal solid wastes requires sludge cakes with a solids concentration of at least 20%, otherwise leachates may increase excessively in the landfill, threatening side slopes stability.

In many places, bulldozers are usually employed for wastes compression and cell implementation. Low solids content sludge cakes may stick to tractor track plates, reducing its compaction capability.

Co-disposal has a lower application rate when compared to monofill rates, ranging from 200–1,600 tonnes/ha, on a dry basis (Malina, 1993). These are merely referential figures, as the sludge/urban waste ratio must be defined as a function of the characteristics of both residues and landfill itself.

9.5.5 Basic design elements

(a) Landfill capacity

Landfill sizing should be calculated based upon the sludge volume to be disposed of during a certain period, which usually ranges between 15 and 20 years. Future

Table 9.6. Example of required landfill volumes (demand factor) depending upon sludge total solids

Solids content in the sludge (%)	Volumetric demand per tonne of dry matter (m^3/tonne dry solids)
15	6.93
20	5.43
25	4.30
40	2.75
90	1.10
Ashes	0.32

Source: Fernandes (1999)

expansions of the leachate drainage system should be anticipated in the design, considering yearly increments in wastewater sludge plus municipal refuse volumes. Sludge volumes should be computed according to the daily production on a dry basis. Moisture content will severely impact transportation, occupied volume and landfill operation, as exemplified in Table 9.6.

High moisture sludge cakes, besides occupying large fill volumes, may also lead to subsidence of the buried volumes due to the significant water losses resultant from landfill leaching.

The required monthly volume can be calculated as follows:

$$V = P.C.F \times 30 \tag{9.3}$$

where:

V = required fill volume for one month sludge cakes (m^3/month)
P = daily sludge production, on a dry basis (tonne/day)
C = soil daily covering factor (usually 1.2 to 1.5)
F = volume demand factor (m^3 of landfill per tonne of sludge cakes on a dry basis)

(b) Impermeabilisation of landfill bed

The earthmoving cutting plan of soil should be impermeable to prevent leaks and groundwater contamination.

Well-compressed clayey soil can reach an acceptable permeability coefficient ($K < 10^{-7}$cm/s) if layer thickness is suitable.

Flexible membrane liners (FML) of various thicknesses are commercially available for non-hazardous wastes, such as wastewater sludges and/or municipal solid wastes. An FML thickness of 1–2 mm is usually considered acceptable.

(c) Stormwater drainage system

Surface drains are intended to detour stormwater and reduce the amount of leaching liquids in the landfill. Its network must be designed according to site topography, avoiding the landfill leachate collection system and soil erosion at discharge point.

Definitive drainage collectors are usually made of open concrete pipes, whereas temporary drain systems, due to the dynamic landfill construction feature, may consist of open corrugated metal pipe, or a riprap channel.

Storm water drainage system must be compatible with the size of the catchment area, top soil permeability, rainfall rate and other site characteristics.

(d) Leachate collection system

Drainage sizing is not a simple task, as leachate flow depends upon a number of intervening factors, mainly local rainfall rate and moisture content of landfilled sludge.

Leachate collection system consists of a small-slope underground ditch, usually excavated in the soil. A porous non-woven geotextile membrane is put along the ditch bottom and large diameter rocks are settled on top. Once the rocks are conveniently placed, the blanket is folded wrapping up the rocks. A layer of coarse sand is then spread over the membrane for further protection against geotextile clogging before sludge is finally applied on top.

(e) Gas collection system

Anaerobic decomposition of the organic matter produces gases (CH_4, CO_2, H_2S and others), which need to be collected to avoid its uncontrolled dispersion.

Gas collection system may consist of perforated pipes, vertically settled, externally surrounded by stones to keep holes free from clogging, and horizontally apart no more than 50 m from each other. They are usually settled over the leaching collection system, facilitating gas circulation.

(f) Leachate treatment

As leachates contain a high concentration of pollutants (Table 9.7), they should not be disposed of before undergoing treatment.

Table 9.7. Typical ranges for leachates constituents

Parameter	Concentration
TOC	100–15,000 mg/L
COD	100–24,000 mg/L
Cd	0.001–0.2 mg/L
Cr	0.01–50 mg/L
Zn	0.01–36 mg/L
Hg	0.0002 0.0011 mg/L
Pb	0.1–10 mg/L
Faecal coliforms	2,400–24,000 MPN/100mL

Source: Malina (1993)

Table 9.8. Support buildings and appurtenances in sludge landfills

Item	Comments
Sentry-box	Intended for controlling admittance. May have logbooks, notepads and documentation regarding truck weighing, if necessary
Scale	Necessary if hauling payment is based on weight carried to site
Isolation distances	The area must be enclosed to avoid foreign personnel admittance. A wire fence or a barbed wire may be provided. A live shrub fence is advised as visual barrier
Shed and workshop	A shed with dimensions compatible with machine and materials routinely used should be provided. This shed can also keep basic tools used in machinery daily maintenance
Office	Located nearby the sentry-box or in another place within the site. The office should keep landfill operational record data, documentation, change rooms and water closets
Internal roads	Access inner roads allow truck traffic to the working front and may change from time to time due to landfill dynamics. Roads shall assure good transit even in rainy days

Biological methods are usually employed for leachate treatment. Conventional stabilisation ponds are not recommended when leachates are highly concentrated. There is no widely accepted solution for leachate treatment systems. One of the difficulties is simply the heterogeneity of the effluent, with a broad variation in composition due to the large range of wastes disposed of in landfill sites. The most used leachate treatment process is the aerobic biological treatment, however, special attention is required while assessing nutrients availability, as nutrient addition may be necessary. Physicochemical treatment is also employed, mainly to improve effluent quality (polishing) and to reduce metals and phosphorus concentration. Other solutions, such as recycling or irrigation, might also be feasible if leachate volumes are not excessively high.

9.5.6 Support buildings

Table 9.8 shows main ancillary buildings and appurtenances usually needed in a landfill site.

9.5.7 Landfill monitoring

The landfill must be monitored throughout its lifetime and for many years after operation is discontinued, since leachates and gases will continue to be produced for over 20 years after its closure.

Water table monitoring is certainly the most important item to be evaluated. This may be done using 30-cm diameter monitoring wells made with PVC or steel lining. Perforation must cease few meters below the water table level and the shaft must have its top end sealed to keep water off outer contamination.

All water inside the pit should be drained off using a portable pump before sample collection takes place. Collecting frequency and parameters to be analysed must be defined in the monitoring plan.

For municipal solid waste landfills, one monitoring well is recommended upstream and three downstream, located at convenient places. Wells location should be performed by an experienced hidrogeologist according to a monitoring plan approved by the local environmental agency.

The monitoring should also include other items, such as gas production and differential settlement control, according to the assigned future use of the area.

9.5.8 Landfill closure

Once the useful volume of a landfill is filled, its lifetime is over and the area can be released for other uses.

The landfill project must consider a closure plan defining the future use of the area. This is very important, as it can orient the operation of the landfill, especially when close to lifetime span, when levels and plans need to be implemented to conform to the expected future use.

Since landfill sites are usually far from the urban perimeter, the future use of the area is normally associated with parks, green areas and sports activities. Housing projects should not be allowed, unless adequate foundation, gas collection system and safety measures are provided.

Example 9.1

From Example 2.1 (Chapter 2), wastewater from 100,000 inhabitants is treated by an anaerobic sludge blanket reactor (UASB). Estimate the area yearly needed for disposal of the dewatered sludge in an exclusive sanitary landfill using both alternatives: narrow (3 m) and large (15 m) trenches. Data from Example 2.1:

- Dewatered sludge solids production: 1,500 kgSS/d
- Daily dewatered sludge volume production: 4.0 m^3/d
- Sludge density: 1,050 kg/m^3

Trench dimensions adopted in this example:

- Trench length: 100.00 m
- Trench depth: 2.50 m

Solution:

(a) Initial information

SS concentration in sludge to be landfilled = (1,500 kg/d) / (4,0 m^3/d)

= 375 kg/m^3

Solids content = (375 kg/m^3) / (1,050 kg/m^3) = 0.36 = 36%

Example 9.1 (Continued)

Trench depth is defined by the water table level and available equipment. Small size bulldozers can deal with 2.50 m depth trenches. This is a popular equipment amongst local government departments as well as contractors. The 2.50 m depth assumes a minimum vertical distance of 1.50 m between the trench bottom and the water table level.

(b) Narrow trenches (3 m)

Volumetric capacity of each trench: 3.00 m × 2.50 m × 100.00 m = 750 m^3

A daily disposal routine requires sludge to be covered at the end of every day shift. Assuming a 25% soil-to-sludge ratio by volume, the trench will be able to store 750 m^3/(1.00 + 0.25) = 600 m^3 of sludge (wet weight). The remaining 150 m^3 is for soil cover volume.

Example 2.1 shows that for a dewatered sludge volume of 4.0 m^3/d and specific weight of 1.05, the total sludge mass (dry solids + water) hauled to the landfill is 4.0 × 1.05 = 4.2 tonne/d. The weight determination is necessary whenever freight is paid by weight instead of volume.

Therefore, 1 m^3 of landfill volume can accept 1.05 tonne of sludge (wet weight). Then:

Yearly sludge production: 4.2 tonne/d × 365 d/year = 1,533 tonne/year.
Yearly sludge volume: 4.0 m^3/d × 365 d/year = 1,460 m^3/year.
Number of required cells: (1,460 m^3)/(600 m^3/cell) = 2.43 cells.

Assuming 10 m between contiguous cells and not computing external border space, which is variable from site to site, and assuming a rectangular shaped terrain, a possible cell arrangement is presented in the following diagram:

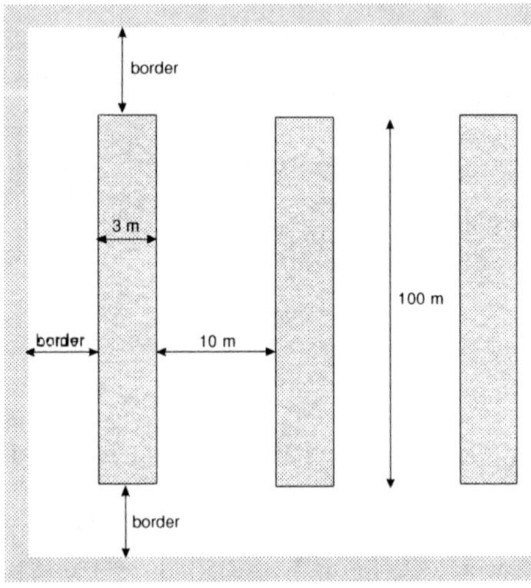

Example 9.1 (Continued)

In a period of one year 2.43 cells are necessary.

The effective area occupied by the cells is:

$$\text{Area} = 2.43 \times 100\ \text{m} \times (3\ \text{m} + 10\ \text{m}) = 3{,}159\ \text{m}^2/\text{year}$$

This area may change if depth and distribution of the cells are rearranged.

(c) Large trenches (15 m)

Volumetric capacity of each trench: $15\ \text{m} \times 2.5\ \text{m} \times 100\ \text{m} = 3{,}750\ \text{m}^3$.

Assuming 25% as the covering coefficient, the useful capacity will be $3{,}750\ \text{m}^3/1.25 = 3{,}000\ \text{m}^3$.

Yearly sludge volume: $1{,}460\ \text{m}^3$/year (calculated in Item b)

Number of cells required: $(1{,}460\ \text{m}^3)/(3{,}000\ \text{m}^3/\text{cell}) = 0.49$ cells

Just 0.49 cell would be sufficient to absorb the yearly sludge production, which is equivalent to saying that one cell shall be enough to absorb the sludge production of approximately 2 years.

Assuming the same spacing between cells, the yearly needed area is:

$$\text{Area} = 0.49 \times 100\ \text{m} \times (15\ \text{m} + 10\ \text{m}) = 1{,}225\ \text{m}^2/\text{year}$$

(d) Comments

In this particular case, considering a 36% total solids sludge cakes, any of the two alternatives could be used. If centrifuges or belt presses had been used for sludge dewatering, solids content would be below 25%, and large trenches would not be recommended, due to their incapability to support traffic vehicles onto sludge layers.

It is interesting to notice from Table 9.6 how significant is the impact of sludge cake solids content in landfill: as solids content increases, the required volume for disposal is substantially reduced. As it can be seen, a 15% total solids sludge demands $6.93\ \text{m}^3$ of useful landfill volume for 1.0 tonne of sludge (dry basis), whereas a 25% solids sludge demands a landfill volume of only $4.30\ \text{m}^3$ per tonne (dry basis).

10

Environmental impact assessment and monitoring of final sludge disposal

A.I. de Lara, C.V. Andreoli, E.S. Pegorini

10.1 INTRODUCTION

Feasible alternatives for final sewage sludge disposal, as stated by Agenda 21, are a worldwide concern. They should focus on adequate waste management and accomplish the following principles: all residues should be minimised, reuse and recycling should be practised whenever possible and remaining residue should be properly disposed.

The primary concern of the selected sludge disposal alternative should be health and environmental protection. Achievement of such goals requires a sound assessment of environmental impacts and risks regarding the selected disposal method, aiming to minimise negative impacts and emphasise the positive ones.

From the early stages of a wastewater treatment plant planning and design, beneficial use or final disposal alternatives for the produced sludge should be considered, along with pertinent technical, economic, operational and environmental aspects of the problem. The entity that generates a residue is responsible for its safe and adequate destination, and this is particularly true for water and sanitation companies.

ISBN: 1 84339 166 X. Published by IWA Publishing, London, UK.

The hazards and indicators of risk associated with sludge disposal methods are discussed in the present chapter, together with the corresponding monitoring programme. It should be emphasised that impacts can be positive or negative, that is, they can add value or depreciate a particular disposal alternative. Chapters 8 and 9 discuss positive and negative impacts, but these are analysed in the present chapter in an integrated view with the main disposal alternatives.

10.2 POTENTIALLY NEGATIVE ENVIRONMENTAL IMPACTS

These impacts may be more or less complex, depending upon the amount of sludge to be disposed of and on the physical, chemical and biological characteristics of the sludge, as well as frequency, duration and extent of the disposal. These factors, amongst others, determine the importance and magnitude of the impacts related with the selected sludge disposal alternative.

The present section analyses negative impacts arising from the following sludge disposal routes:

- ocean disposal
- incineration
- sanitary landfill
- landfarming
- beneficial land application

(a) Ocean disposal

Marine disposal is a forbidden practice in most countries, since it is potentially able to produce negative impacts to the marine environment. Sewage sludge may bring pathogens, toxic organic compounds and metals. Some of these may settle to the bottom of the sea, contributing to alter the benthic community, leading to death of sensitive species, or bioaccumulating metals and toxic compounds in the trophic chain, finally reaching human beings through ingestion of contaminated fish and mussels. Moreover, plankton growth and resulting increase in dissolved oxygen consumption is furthered by nutrients in sludge.

According to Loehr (1981), estimates about sludge disposal impacts on oceans are not consolidated. There is not enough information about the residue dispersion dynamics in seawater, organic matter decomposition rate, transport of toxic elements and pathogenic organisms, composition of benthic fauna and production of aquatic wildlife in coastal areas. These arguments show that marine disposal is an alternative whose environmental effects cannot be easily measured and controlled.

(b) Incineration

Incineration is not considered as a final disposal practice by several authors, since this process generates ashes as residue, which must be adequately disposed of.

Table 10.1. Potential air pollution due to sludge incineration

Pollutant source in sludge	Pollutant
Volatile solids	Organics (PCB and others) Odour Hydrocarbons
Ashes	Suspension of particulates Metals
Burning process	Carbon monoxide Partially oxidised hydrocarbons Sulphur oxides (SO_2, SO_3) Nitrogen oxides (NO_x)
Ashes handling	Pollutants in ashes
Auxiliary fuel incineration	Ash pollutants Pollutants from combustion process

Depending on the sludge characteristics, 10 to 30% of the total dry solids are transformed into ashes, which are commonly landfilled. Ashes landfilling are an additional impact related to incineration, since compounds not eliminated by thermal destruction, as metals, are concentrated in the ashes.

The main impact of sludge incineration is air pollution through emission of gases, particulates and odour (see Table 10.1). The severity of this impact may be higher if the system is not properly operated. Neighbouring communities may face health problems due to atmospheric pollution and are directly affected by the aesthetic aspects.

(c) Landfill

Like any other form of wastewater sludge disposal, sludge monofills or co-disposed with municipal solid wastes require adequate site selection.

The main impact of landfills is on surface or groundwater that might become contaminated by leaching liquids carrying nitrates, metals, organic compounds and pathogenic microorganisms. As a result of the anaerobic stabilisation process carried out in landfills, gases are produced, which need to be exhausted and controlled.

Environmental impacts from landfilling wastewater sludges may decrease if the site is well located and protected, leachate treatment is provided, gases are properly handled and the landfill is efficiently managed and operated.

(d) Landfarming

Landfarming is an aerobic treatment of the biodegradable organic matter that takes place on the upper soil layer. Sludge, site, soil, climate and biological activity interact in a complex dynamic system in which the component properties modify

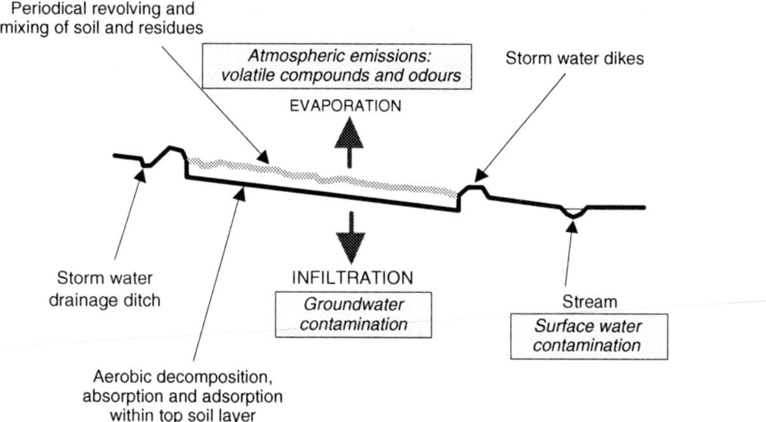

Figure 10.1. Schematics of landfarming and possible associated environmental impacts (adapted from CETESB, 1985)

with time. Since it is an open system, wrong planning and management may cause contamination of water sources, food and soil itself (Figure 10.1).

Land treatment of wastewater sludges are usually destined for environmentally hazardous residues with high concentration of hardly decomposable pollutants which, when successively applied, will accumulate on soils. These substances may then render the landfarming areas impracticable for any further use.

(e) Beneficial land application

Land application of sludge may alter the physical, chemical and biological soil characteristics. Some changes are beneficial, whilst others may be undesirable. Positive impacts are related to organic matter and nutrients added to soil, fostering its physical and chemical properties and microbial activity.

Negative impacts are consequences of (a) accumulation of toxic elements, mainly metals, organics and pathogens, on soil; (b) leaching of constituents resulting from sludge decomposition, mainly nitrates; (c) storm run-off flows, contaminating nearby areas and water bodies; (d) volatilisation of compounds that, although less significant, may lead to foul odours and vector attraction (Figure 10.2).

The severity of those negative impacts depends on the disposal technique. Land reclamation and agricultural recycling, discussed below, are two possible methods of land application.

Land reclamation. Large amounts of sludge are employed in the recovery of degraded areas, either those resulting from inadequate agricultural handling or

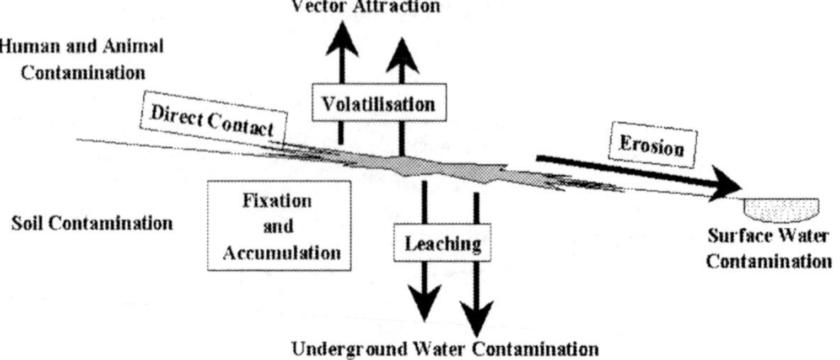

Figure 10.2. Direct impacts of sludge disposal on soil

from extractive activities, increasing the amounts of undesirable elements in soil, depending upon sludge characteristics. When applying high rates of sludge on land, careful analysis of imbalances that may occur between soil nutrients and leached nitrates is required.

Degraded areas are not structurally defined, typically presenting top and sub-surface layers mixed up, and having direct influence of climatic variations which may increase the susceptibility to erosion and leaching. As public access to those often-distant areas is restricted, odours and vectors are less significant items. Should erosion be a serious consideration on a particular degraded area, application of high rates of sludge is inappropriate, because this may lead to deterioration of run-off quality.

Agricultural land application. The main impacts of agricultural recycling are associated with the contamination hazards by toxic elements and pathogens, since both may affect environmental quality and public health. Applied rates should be based on crop nitrogen demand to avoid leaching and nitrates to the water table. Especially in case of lime-treated sludges, pH control to reach the desired level is important, together with nutrients balance in sites with continuous application. These risks are minimised through careful selection of the application sites, considering sludge, soil and physical characteristics, aiming to control:

- toxic elements and pathogenic organisms (accumulation and fixation) input
- natural dispersion mechanisms (storm run-off and leaching)
- indirect contamination (population and water-bodies vicinity, animal grazing and edible crops contamination)
- nutrients balance

10.3 MONITORING INDICATORS AND PARAMETERS

Accomplishment of an efficient monitoring relies on suitable environmental indicators. Each sludge disposal method has an appropriate indicator for impact assessment of the selected alternative. For instance, monitoring water quality may

Table 10.2. Main indicators related with impacts of sewage sludge disposal

Impact	Indicators
Water pollution	• changes in water quality • concentration of contaminants (toxic compounds and pathogens) • bioindicator species of environmental quality
Air pollution	• presence of gases and toxic substances • presence of particulates • odours
Soil pollution	• changes in physical, chemical and biological soil properties • concentration of contaminants (toxic compounds and pathogens)
Transmission of diseases	• pathogens density in soil • vectors attractiveness on application site (rodents and insects) • pathogenic organisms and toxic compounds concentration in crops
Food chain contamination	• concentration of contaminants in water, soil and crops • disturbances in wildlife communities • bioindicator species
Aesthetic and social problems	• acceptability in disposal area neighbourhood • consumers and producers acceptability of goods from sludge-amended areas • properties depreciation near sludge disposal sites

be more suitable and relevant for a particular disposal alternative than odour emission. Obviously, both must be monitored, but the impact on water quality resources has greater magnitude and importance than foul odours, since it potentially affects more people. Table 10.2 presents the main indicators related with the impacts of sewage sludge disposal alternatives.

Analytical parameters must be defined for each indicator to provide quantitative and qualitative data in the monitoring process that may lead to conclusions on the practice being carried out for sludge disposal. The selection of proper indicators and monitoring parameters depends on the adopted disposal alternative, sludge characteristics, monitoring objectives and requirements of local environmental legislation. Parameters used for water, soil and crop monitoring of sludge disposal sites are shown in Table 10.3.

Microbial soil communities can also be employed as monitoring parameters. According to Lambais and Souza (2000), both microbial soil biomass and its metabolic activities which can change the microbial communities may be affected by potentially pollutant agents, implying that such parameters may be useful for environmental impacts assessment and soil quality monitoring. Cardoso and Neto (2000) suggest the following parameters: CO_2 release, carbon biomass, enzymatic activity, counting of nitrogen-fixing microorganisms and mineralisation of nitrogen.

Table 10.3. Typical physical and chemical parameters for sludge disposal sites monitoring

Source	Parameters
Groundwater	pH, conductivity, total hardness, total dissolved solids, sulphates, total organic carbon, nitrate, nitrogen, total phosphorus, surfactants, metals or trace organics selected as necessary, indicator organisms
Surface water	Faecal coliforms, total phosphorus, total Kjeldahl nitrogen, dissolved oxygen, BOD, temperature, pH, suspended solids
Soil	Nitrates, total nitrogen, phosphorus, pH, conductivity, organic carbon, exchangeable cations (calcium, magnesium, potassium, sodium), metals (lead, mercury, chromium, cadmium, copper, nickel, zinc), CEC (Cation Exchange Capacity), texture, other components[1]
Crop	Metals (lead, mercury, chromium, cadmium, copper, nickel, zinc), macronutrients (NPK), other components[1]

[1] Other components, such as As, Fe, Mo, Se, PCBs, DDT and Dieldrin, must be analysed only if there are reasons to believe that significant quantities may be present in the sludge.
Source: Adapted from Granato and Pietz (1992)

10.4 MONITORING PLAN

Monitoring plans are useful instruments to control and assess the efficacy of the entire sludge disposal operation. They allow (a) to control and supervise impacts, (b) to follow the implementation and execution of the control measures, (c) to adjust, calibrate and validate models and parameters, and (d) to serve as reference for future studies monitoring propositions.

Monitoring responsibilities must be defined among the various parties involved: environmental agency, entrepreneur, other governmental and departmental agencies and the affected community.

Monitoring efficacy will depend on a plan identifying impacts, indicators and parameters, sampling frequencies, sampling points and analytical methods, leading to comparative and publishable results. The following elements are necessary while preparing a monitoring plan:

Monitoring goals. Clear and objective statement on monitoring purposes as a function of the selected final disposal alternative and possible related impacts.

Review of existing data. Encompasses a description of the selected alternative, characteristics of the disposal area(s), evaluation of the impacts and sludge characteristics. All information gathered on the final disposal site prior to process start-up may suit as future reference for comparison purposes. These tests prior to sludge application should be undertaken on the possible sources of concentration of contaminants (air, water, soil).

Definition of impacts. Relates to the potential consequences (impacts) the proposed activity may have upon the environment.

Selection of impact indicators. There is no list of applicable parameters for all cases. The legal requirements established for different kinds of wastes disposal in each region may serve as groundwork for choosing parameters. Existent constituents in sludge which may be present in concentrations that may deteriorate environment quality should be necessarily monitored.

Critical levels. Environmental critical levels allow the interpretation and assessment of the impact intensity, and may be either single figures or range limits.

Analytical and data collection methodology. Selection of laboratory sampling methods and procedures should consider the capability of existent laboratories near the disposal area, the parameters to be analysed and the size of the total disposal area. Sampling methodology must guarantee representativeness of the indicator, and the analytical procedures must be defined and calibrated to produce reliable data within a pre-defined accuracy.

Sampling points. Data should be collected where the occurrence of an impact is more likely to occur, allowing characterisation of the areas with lower or higher alterations.

Monitoring frequency. Sampling frequency of the selected parameters should be defined for both the sludge and the disposal area, and should allow identification of critical periods within seasonal variations.

USA sludge regulation – USEPA 40CFR Part 503 (EPA, 1993) requires that land applied **sludge** be monitored for metals, density of pathogens and parameters indicating vector attraction reduction. Frequency of sampling is dependent on the quantity of biosolids applied during one year (Table 10.4). Table 10.5 presents the monitoring frequency requirements established for the Brazilian State of Paraná, which has a large programme of biosolids recycling.

Monitoring frequency for the **sludge disposal areas** must be determined through assessment of the effects of the application to structure a database with the information gathered from each application site. A sampling network should be established on the application site and surroundings, defining sampling points

Table 10.4. Monitoring frequency for pollutants, pathogen density and vector attraction reduction (USEPA 40CFR Part 503)

Amount of biosolids land applied (tonne/year) – dry basis	Frequency
0–290	Once a year
290–1,500	Four times a year
1,500–15,000	Six times a year
≥15,000	Once a month

Source: EPA (1993)

Table 10.5. Sampling frequency for characterisation of biosolids for agriculture recycling (Paraná State, Brazil)

Biosolid land application (tonne/year) – dry basis	Frequency
<60	Once a year (prior to the highest demand harvest)
60–240	Every 6 months (once before summer harvest and another before winter harvest)
>240	Every biosolid lot of 240 tonne (dry matter) or every semester (whichever comes first)

Source: Fernandes *et al.* (1999)

Table 10.6. Monitoring frequency for wastewater sludge monofills and dedicated land disposal (DLD) sites

	Sludge		Groundwater[1]		Soil[2]	
Parameter	Unit	Frequency	Unit	Frequency	Unit	Frequency
Total nitrogen	mg/kg	Monthly	mg/L	quarterly	mg/kg	quarterly
Nitrate nitrogen	mg/kg	Monthly	mg/L	quarterly	mg/kg	quarterly
Ammonia nitrogen	mg/kg	Monthly	mg/L	quarterly	mg/kg	quarterly
Phosphorus	mg/kg	Quarterly	mg/L	quarterly	mg/kg	2/month
Potassium	mg/kg	Quarterly	mg/L	quarterly	mg/kg	2/month
Cadmium	mg/kg	Quarterly	mg/L	quarterly	mg/kg	2/month
Lead	mg/kg	Quarterly	mg/L	quarterly	mg/kg	2/month
Zinc	mg/kg	Quarterly	mg/L	quarterly	mg/kg	2/month
Copper	mg/kg	Quarterly	mg/L	quarterly	mg/kg	2/month
Nickel	mg/kg	Quarterly	mg/L	quarterly	mg/kg	2/month
pH	–	Monthly	–	quarterly	–	quarterly
PCB	mg/kg	Yearly	mg/L	yearly	mg/kg	yearly
Water level	–	–	Meter	quarterly	–	–
CEC	–	–	–	–	meq/100g	quarterly

[1] One well each 20 ha of DLD
[2] One sample at 15 cm, 45 cm and 75 cm for each 8 ha of DLD
Source: Griffin et al. (1992)

including all possible media (air, water, soil, and crops), depending on the selected disposal alternative.

Wastewater sludge monofills and dedicated land disposal (DLD) sites might be monitored as advised by Griffin *et al.* (1992) (Table 10.6). The authors still recommend monthly monitoring of gas collection points in landfills with a portable gas detector, increased to weekly verifications if high levels of gases are identified.

For reclamation of **degraded areas**, Gschwind and Pietz (1992) present a minimum list of parameters, which should be included in routine water, soil and vegetation laboratory analyses (Table 10.7).

Data tabulation, analysis and evaluation. The analytical results could lead to a database with detailed information from the sludge disposal site, supported by a geo-referenced system.

Table 10.7. Minimum sampling procedure in degraded areas

Sample	Procedure
Water	• Collect at least three samples from every groundwater well and lysimeter station, prior to sludge application • Collect monthly water samples, after the application of sludge, during one year • For samples prior to sludge application and for those corresponding to the first three months after application, pH, Cl, NO_3-N, NH_4-N, Org-N, Fe, Al, Mn, Cu, Cr, Co, Pb, Cd, Ni, Zn and faecal coliforms should be analysed • From the 4th to the 11th month after application, only pH, NO_3-N, NH_4-N, Zn, Cu, Pb, Co, Ni, Cd, Cr and faecal coliforms should be analysed • In the 12th month after application, pH, Cl, NO_3-N, NH_4-N, Org-N, Fe, Al, Mn, Cu, Cr, Co, Pb, Cd, Ni, Zn and faecal coliforms should be analysed • Water sampling may end after one year, unless if ¾ of the data indicate that the process should continue. If more sampling is needed, the samples should be collected quarterly until sufficient data is gathered to allow conclusions • Monitoring of wells should continue after the first year to corroborate the data acquired in the last data collection
Soil	• Soil samples should be collected before sludge application. Surface samples should be collected at several points and analysed for pH, verifying whether liming is needed to raise pH level up to 6.5. Also soil CEC (Cation Exchange Capacity) should be determined. Samples from soil profile must be collected from pits excavated for lysimeters at 0–15 cm, 15–30 cm, 30–60 cm and 60–90 cm depths • One year after sludge application, soil samples should again be collected at 0–15 cm, 15–30 cm, 30–60 cm depths • All soil samples must be analysed for pH, P, Ca, Mg, K, Na, Fe, Al, Mn, Cu, Zn, Cr, Co, Pb, Cd, Ni and N Kjeldahl • Two years after application, the topsoil should once more be analysed for pH to check whether it still remains bellow 6.5
Vegetation	• Foliar samples should be analysed by the end of the growing season, after biosolid application. Separate samples from each planted species must be collected and analysed for N, P, K, Ca, Mg, Fe, Al, Mn, Cu, Zn, Cr, Co, Pb, Cd and Ni • For sown sites in fall seasons, vegetation samples should be collected at the end of the next season

Source: Gschwind and Pietz (1992)

Data analysis is essential in the decision-making process of whether a particular sludge disposal site should continue to be used, and provides useful input related to corrective measures that might be taken to achieve the desired programme goals. Furthermore, the analysis should also contribute to a better assessment of the parameters effectiveness and suitability of the analytical methods being used.

Maximum allowable concentrations for pollutants are useful as references for data interpretation. However, it should be borne in mind that specific legislations reflect local or regional characteristics, and may not be widely applicable in every country or region. The best approach would be for each region to develop its own studies aiming at soil characteristics identification to establish proper legal parameter values.

Reports. The entity in charge of the final sludge disposal must establish a sound relationship with the community and environmental agencies, especially those in the surroundings of the sludge disposal area. Periodical reports should be sent to environmental agencies, showing clearly and objectively the interpreted monitoring results. This helps to build a historical database, open for public consultation. The reports and analytical results should be filed in the sludge-generating site, for occasional inspection by environmental protection agencies.

Information to population. The involved community should have access to any relevant information about environmental impacts such as to guarantee the transparency of the process.

Final remarks. Monitoring should be viewed as an integral part of the final sludge disposal process, since every alternative may potentially affect air, soil, water and crop quality.

The joint participation of the community and environmental agencies in all stages of the process, from the conception of the disposal project to the execution of its monitoring, allows improvements and control over the process, minimising possible negative impacts from the selected sludge disposal alternatives.

A monitoring plan is a dynamic instrument within the process, and in constant improvement from the very beginning of its implementation, because it is fed by the analysis of the results obtained and moves forward by the continuous research progress on sludge beneficial uses and disposal.

References

ABNT (1989). *Projeto de estações de tratamento de esgotos* (in Portuguese), NB 570, Associação Brasileira de Normas Técnicas.

ABNT (1992) NBR 8,419 (in Portuguese), Associação Brasileira de Normas Técnicas, Rio de Janeiro.

ABNT (1997) *Tratamento no solo (landfarming)* (in Portuguese). NBR 13,894, Associação Brasileira de Normas Técnicas, Rio de Janeiro.

ADEME (1998) *Connaissance et maîtrise des aspects sanitaires de l'épandage des boues d'épuration des collectivités locales.*

Agrodevelopment, S.A. (1995) *Elaboração do plano piloto para gestão e reciclagem agrícola de lodo da ETE-Belém* (in Portuguese). Harry, J. Convênio de cooperação técnica SANEPAR – Governo Francês.

Aisse, M.M., Van Haandel, A.C., Von Sperling, M., Campos, J.R., Coraucci Filho, B., Alem Sobrinho, P. (1999) Tratamento final de lodo gerado em reactores anaeróbios. In *Tratamento de esgotos sanitários por processo anaeróbio e disposição controlada no solo* (coord. J.R. Campos) pp. 271–300 (in Portuguese), PROSAB.

Andreoli, C.V., Barreto, C.G., Bonnet, B.R.P. (1994) Tratamento e disposição final do lodo de esgoto no Paraná (in Portuguese). *Sanare*, **1**(1), 10–15.

Andreoli, C.V., Pegorini, E.S. (1999) Distribution plans and legal permission requirements for land application of biosolids in Paraná, Brazil. In *IAWQ Specialized Conference on Disposal and Utilization of Sewage Sludge*, 71–78.

Andreoli, C.V., Lara, A.I., Fernandes, F. (1999) *Reciclagem de biossólidos – Transformando problemas em soluções* (ed. Sanepar) (in Portuguese).

Andreoli, C.V., Von Sperling M., Fernandes, F. (2001) Lodo de esgotos: tratamento e disposição final. *UFMG – Sanepar* (in Portugese) 484 pp.

Ayres, R.M., Lee D.L., Mara, D.D., Silva, S.A. (1994) The accumulation, distribution and viability of human parasitic nematode eggs in the sludge of a primary facultative waste stabilization pond. *Trans. R. Soc. Trop. Med. Hyg.* **87**, 256–258.

Berron, P. (1984) Valorisation agricole des boues d'épuration: aspects micro biologiques, *ISM.* **11**, 549–556.

Brown, L. (1993) *Qualidade de vida – Salve o planeta.* São Paulo: (ed. Globo) (in Portuguese).

Cardoso, E.J.B.N., Fortes Neto, P. (2000) Aplicabilidade do biossólido em plantações florestais: III Alterações microbianas no solo. In *Impacto ambiental do uso agrícola do lodo de esgoto*. BETTIOL, W. e CAMARGO, O.A. (orgs). Jaguariúna, SP: EMBRAPA Meio Ambiente, p. 197–202 (in Portuguese).

CEMAGREF (1990) *Les systèmes de traitment de boues des stations d'épuration des petites collectivités*. Documentation technique, FNDAE **9**, 85. Centre National du machinisme agricole du génie rural des eaux et des forêts, Paris.

CETESB (1985) *Resíduos sólidos industriais*. Série Atas da Cetesb, São Paulo: CETESB/ASCETESB (in Portuguese).

Chernicharo, C.A.L. (1997) *Princípios do tratamento biológico de águas residuárias. Vol. 5. Reactores anaeróbios*. DESA-UFMG. 245 pp (in Portuguese).

CIWEM (The Chartered Institution of Water and Environmental Management) (1996) Sewage sludge – Utilization and disposal. In *Handbook of UK Wastewater Practice*, 110 pp.

CIWEM (The Chartered Institution of Water and Environmental Management) (1999) Sewage sludge – Conditioning, dewatering, thermal drying and incineration. In *Handbook of UK Wastewater Practice*, 110 pp.

Crites, R., Tchobanoglous, G. (2000) *Tratamiento de aguas residuales en pequeñas poblaciones* (in Spanish), McGraw-Hill, Colombia.

Da Silva, S.M.C.P., Fernandes, F., Soccol, V.T., Morita, D.M. Principais contaminantes do lodo (2001) In *Princípios do tratamento biológico de águas residuárias. Vol. 6. Lodo de esgotos. Tratamento e disposição final* (in Portuguese) (eds. C.V. Andreoli, M. Von Sperling, M., F. Fernandes), p. 484, Departamento de Engenharia Sanitária e Ambiental – UFMG. Companhia de Saneamento do Paraná – SANEPAR.

EPA (1979) *Process Design Manual – Sludge Treatment and Disposal*. U.S. Environmental Protection Agency.

EPA (1983) *Land Application of Municipal Sludge – Process Design Manual*. EPA/625/1–83–016, United States Environmental Protection Agency, Cincinnati.

EPA (1987) *Design Manual – Dewatering Municipal Wastewater Sludges*. EPA/625/1–87/014, United States Environmental Protection Agency.

EPA (1990) *Autothermal Thermophilic Aerobic Digestion of Municipal Wastewater Sludge*. EPA/625/10–90/007, United States Environmental Protection Agency.

EPA (1992) *Control of Pathogens and Vector Attraction in Sewage Sludge Under 40 CFR part 503*. Office of Water, Office of Science and Technology Sludge Risk Assessment Branch. Washington, DC.

EPA (1992) *Control of Pathogens and Vector Attraction in Sewage Sludge*. EPA/625/R–92/013, United States Environmental Protection Agency.

EPA (1993) *Land Application of Biosolids – Process Design Manual*. United States Environmental Protection Agency, Cincinnatti.

EPA (1994) *A Plain English Guide to the EPA Part 503 Biosolids Rule*. EPA/832/R–93/003, United States Environmental Protection Agency.

EPA (1995) *A Guide to the Biosolids Risk Assessments for the EPA Part 503 Rule*. EPA/832/B–93/005, United States Environmental Protection Agency.

EPA (1999) *Biosolids Generation, Use and Disposal in the United States*. EPA/530/R–99/009, United States Environmental Protection Agency.

European Environmental Agency (1997) Sludge treatment and disposal. *Environmental Issues Series*. p.52.

Feix I., Wiart J. (1998) *Les agents pathogènes. Connaissance et maîtrise des aspects sanitaires de l'épandage des boues d'épuration des collectivités locales*, ADEME, p. 75.

Fernandes, F. (1999) *Relatório PROSAB 01, Tema 4*. Universidade Estadual de Londrina (in Portuguese).

Fernandes, F., Lara, A.I., Andreoli, C.V., Pegorini, E.S. (org.) (1999) Normatização para a reciclagem agrícola do lodo de esgoto. In *Reciclagem de biossólidos: transformando problemas em soluções*, (coord. C.V. Andreoli, A.I. Lara, F. Fernandes), pp. 263–291, Curitiba. Sanepar/Finep. (in Portuguese).

Fernandes, F., Silva, S.M.C.P. (1999) *Manual prático para compostagem de biossólidos*. PROSAB – Programa de Pesquisa em Saneamento Básico – ABES, Rio de Janeiro, 84 p (in Portuguese).

Fernandes, F. (2000) Estabilização e higienização de biossólidos. In *Impacto ambiental do uso agrícola do lodo de esgotos*. (ed. Embrapa Meio Ambiente) (in Portuguese).

Ferreira Neto, J.G. (1999) *Fluxograma ETE Brasília Norte e ETE Brasília Sul* (in Portuguese).

Gaspard P.G., Schwartzbrod J. (1995) Helminth eggs in wastewater: quantification technique. *Wat. Sci. Technol.* **31**(5/6), 443–446.

Gaspard P.G., Wiart J., Schwartzbrod J. (1996) A method for assessing the viability of nematode eggs in sludge. *Environ. Technol.* **17**, 415–420.

Gómez, J.S. (1998) *Impacto ambiental em rellenos sanitarios*. Associación Mexicana para el Control de los Resíduos Sólidos Peligrosos. (in Spanish).

Gonçalves, R. (1996) Sistemas aeróbios com biofilmes. In *Seminário internacional. Tendências no tratamento simplificado de águas residuárias domésticas e industriais*. (Anais). Belo Horizonte (**6/8**), 128–143 (in Portuguese).

Gonçalves, R. (coord) (1999) *Gerenciamento do lodo de lagoas de estabilização não mecanizadas*. PROSAB. (in Portuguese).

Granato, T.C., Pietz, R.I. (1992) Sludge application to dedicated beneficial use sites. In *Municipal Sewage Sludge Management: Processing, Utilization and Disposal*, (eds. C. Lue-Hing, D.R. Zens and R. Kuchenrither) Water Quality Management Library, **4**, 417–454.

Griffin, R.A. Lue-Hing, C., Sieger, R.B., Uhte, W.R., Zenz, D. (1992) Municipal sewage sludge management at dedicated land disposal sites and landfills. In *Municipal sewage sludge management: processing, utilization and disposal*, (eds. C. Lue-Hing, D.R. Zens and R. Kuchenrither) Water Quality Management Library, **4**, 299–376.

Gschwind, J., Pietz, R.I. (1992) Application of municipal sewage sludge to soil reclamation sites. In *Municipal Sewage Sludge Management: Processing, Utilization and Disposal* (eds. C. Lue-Hing, D.R. Zens and R. Kuchenrither). Water Quality Management Library, **4**, 455–478.

Hall, J.E. (1998) Standardizing and the management of biosolids – The international experience. In *Anais do I Seminário sobre Gerenciamento de Biossólidos no Mercosul*, Curitiba-PR, pp. 113–122.

Helou, L.C. (2000) *Otimização de estações de tratamento de esgotos convencionais por lodos ativados com aproveitamento dos efluentes para reúso*. São Paulo, out/2000. Tese (doutorado) – Escola Politécnica, Universidade de São Paulo. 265 p (in Portuguese).

Hess, M.L. (1973) *Lagoas de estabilização*. Capítulo 6: lagoas anaeróbias. CETESB, São Paulo, 2nd edn (in Portuguese).

Imhoff, K. (1966) *Manual de tratamento de águas residuárias*. Editora da Universidade de São Paulo (in Portuguese).

JICA (1993) *Textbook for the Group Training Course in Sewage Works Engineering*, vol. III.

Jordão, E.P., Pessoa, C.A. (1995) *Tratamento de esgotos domésticos*. ABES, 3rd edn, p. 683 (in Portuguese).

Kiehl, E.J. (1985) *Fertilizantes orgânicos*. Piracicaba: Agronômica Ceres (in Portuguese).

Lambais, M.R., Souza, A.G. (2000) *Impacto de biossólidos nas comunidades microbianas do solo*. In: BETTIOL, W., CAMARGO, O.A. (orgs.). Jaguariúna, SP: EMBRAPA Meio Ambiente, 2000. pp. 269–280 (in Portuguese).

Lashof, D.A., Lipar, D.A. (1989) *Policy Options for Stabilizing Global Climate*. Draft report to congress. EPA, Washington, DC.

Loehr, R.C. Environmental impacts of sewage sludge disposal. In *Sludge and Its Ultimate Disposal* (ed. J.A. Borchardt. Ann Arbor Science. 179–210.

Luduvice, M. (2000) Experiência da Companhia de Saneamento do Distrito Federal na reciclagem de biossólido. In: *Impacto ambiental do uso agrícola do lodo de esgoto.* ed. BETTIOL e CAMARGO. Jaguariúna, SP: EMBRAPA Meio Ambiente. 312 p (in Portuguese).

Malina, J.F. (1993a) Sludge treatment and disposal. In *II Seminário de Transferência de Tecnologia.* ABES, Rio de Janeiro.

Malina, J.F. (1993b) Effectiveness of municipal sludge treatment processes in eliminating indicator organisms. In *Joint Residuals Conference*, AWWA/WEF, Arizona.

Medeiros, M.L.B., Thomaz Soccol, V., Castro, E.A., Toledo, E.B., Borges, J., Paulino, R., Andraus S. (1999) Aspectos Sanitários. In *Reciclagem de biossólidos. Transformando problemas em soluções.* pp. 120–179 (in Portuguese).

Melo, W.J., Marques, M.O. (2000) Potencial do lodo de esgoto como fonte de nutrientes para as plantas. In *Impacto ambiental do uso agrícola do lodo de esgoto.* Bettiol, W. e Camargo, O.A. (coord.), Jaguariúna: EMBRAPA Meio Ambiente, pp. 109–141 (in Portuguese).

Melo, W.J., Marques, M.O., Santiago, G. (1993) Efeito de doses de lodo de esgoto sobre a matéria orgânica e a CTC de um latossolo cultivado com cana. In *Congresso Brasileiro de Ciência do Solo* (24.:1993: Goiânia). Anais . . . Brasília: SBCS, vol. 2, pp. 253–254 (in Portuguese).

Metcalf and Eddy (1991) *Wastewater Engineering: Treatment, Disposal, Reuse*, Metcalf and Eddy Inc., Mac Graw Hill, New York.

Morita, D.M. (1993) Tratabilidade de águas residuárias contendo poluentes perigosos-estudo de caso. Tese (doutorado), Escola Politécnica, Universidade de São Paulo, (in Portuguese).

Muller, P.S.G. (1998) Caracterização físico-química, microbiológica e estudos de desidratação em leito de secagem de lagoas de estabilização. Dissertação de mestrado, UFES, Vitória-ES (in Portuguese).

Nogueira, R., Santos, H.F. (1995) *Cobertura final de células de lodo com solo – Um ensaio realizado* – III Simpósio sobre barragens de rejeitos e disposição de resíduos – REGEO'95, Ouro Preto, MG (in Portuguese).

Oorschot, R.Van, Waal, D., Semple, L. (2000) Options for beneficial reuse of biosolids in Victoria. *Wat. Sci. Technol.* **41**(8), 115–122.

Outwater, A.B. (1994) *Reuse of Sludge and Minor Wastewater Residuals.* S.1.: Lewis Publishers.

Passamani, F.R.F., Motta, J.S., Figueiredo, Gonçalves, R.F. (2000) Pasteurização do lodo de um reator UASB na remoção de coliformes fecais e ovos de helmintos. In *I Seminário Nacional de Microbiologia Aplicada ao Saneamento*, pp. 251, 5–7 Junho 2000, Vitória, Espírito Santo. (in Portuguese).

Passamani, F.R.F., Gonçalves, R.F. (2000) Higienização de lodos de esgotos. In *Gerenciamento de lodos de lagoas de estabilização não-mecanizadas.* Ed. ABES – PROSAB (in Portuguese).

Qasim, S.R. (1985) *Wastewater Treatment Plants: Planning, Design and Operation*, Holt, Rinehart and Winston.

Raij, B. van (1998) Uso agrícola de biossólidos. In *I Seminário sobre Gerenciamento de Biossólidos do Mercosul* (anais), Curitiba: SANEPAR/ABES pp. 147–151 (in Portuguese).

SANEPAR (Companhia de Saneamento do Paraná) (1997) *Manual técnico para utilização agrícola do lodo de esgoto no Paraná.* Curitiba: SANEPAR (in Portuguese).

SANEPAR (Companhia de Saneamento do Paraná) (1997) *Reciclagem agrícola do lodo de esgoto: estudo preliminar para definição de critérios para uso agronômico e de parâmetros para normatização ambiental e sanitária.* Curitiba: SANEPAR (in Portuguese).

SANEPAR (Companhia de Saneamento do Paraná) (1999) *Uso e manejo do lodo de esgotos na agricultura.* ed. PROSAB/FINEP (in Portuguese).

Santos, H.F. (1979) *Aplicação do lodo de estações de tratamento de esgotos em solos agrícolas*, Revista DAE, 122, pp. 31–48 (in Portuguese).

Santos, H.F. (1996) Uso agrícola do lodo das estações de tratamento de esgotos sanitários (ETE's): Subsídios para elaboração de uma norma brasileira. Dissertação para obtenção do grau de mestre em Saneamento Ambiental, Universidade Mackenzie, SP São Paulo (in Portuguese).

Sapia, P.M.A. (2000) Proposta de critérios de recebimento de efluentes não-domésticos para o sistema público de esgoto da Região Metropolitana de São Paulo. Dissertação (mestrado), Escola Politécnica, Universidade de São Paulo (in Portuguese).

Schwartzbrod, J., Mathieu, C., Thevenot, M.T., Baradel, J.M., Schwartzbrod, L. (1990) Sludge parasitological contamination. In *Treatment and use of sewage sludge and liquid agricultural wastes* (ed. L'Hermitte). Congrès d'Athenes (GR), 1–4 Oct 1990, Elsevier Appl. Sci.

Silva, J.E., Dimas, V.S.R., Sharma, R.D. (2000) Alternativa agronômica para o biossólido: a experiência de Brasília. In *Impacto ambiental do uso agrícola do lodo de esgoto*. Bettiol, W. e Camargo, O.A. (coord.), Jaguariúna: EMBRAPA Meio Ambiente, pp. 143–162 (in Portuguese).

Souza, M.L.P., Andreoli, C.V., Pauletti, V. (1994) Desenvolvimento de um sistema de classificação de terras para disposição final do lodo de esgoto. In *Simpósio Luso-Brasileiro de Engenharia Sanitária e Ambiental*, vol.1, tomo 1, pp. 403–419 (in Portuguese).

Thomaz Soccol, V. (2000) Riscos de contaminação do agrossistema com parasitos pelo uso de lodo de esgoto. In *Impacto ambiental do uso agrícola do lodo de esgoto*. ed. BETTIOL e CAMARGO. Jaguariúna, SP: EMBRAPA Meio Ambiente. 312 p.

Thomaz Soccol, V., Paulino, R.C., Castro, E.A. (1997) Helminth eggs viability in sewage and biosolids sludge in Curitiba, Parana, Brazil. *Brazilian Arch. of Biol and Technol.* **40**(4), 829–836.

Thomaz Soccol, V., Paulino, R.C., Castro, E.A. (2000) Metodologia para análise parasitológica em lodo de esgoto. In *Manual de métodos para análises microbiológicas e parasitológicas em reciclagem agrícola de lodo de esgoto*. Andreoli and Bonnet, Curitiba, 2nd edn. rev e ampl. 28–41 (in Portuguese).

Tsutya, M.T. (2000) Alternativas de disposição final de biossólidos gerados em estações de tratamento de esgotos. In *Impacto ambiental do uso agrícola do lodo de esgoto*. BETTIOL, W. e CAMARGO, O.A. (coord.), Jaguariúna: EMBRAPA Meio Ambiente, p. 69–105 (in Portuguese).

UEL (1999) *Manual prático para compostagem de biossólidos*. ed. PROSAB/FINEP (in Portuguese).

Van Haandel, A.C., Lettinga, G. (1994) *Anaerobic Sewage Treatment: A Practical Guide for Regions with a Hot Climate*, John Wiley and Sons.

WEF (Water Environment Federation) (1992) Sludge incineration: Thermal destruction of residues. *Manual of Practice*. FD–19.

WEF (Water Environment Federation) (1996) Operation of municipal wastewater treatment plants. *Manual of Practice.* 11.

WEF/ASCE (Water Environment Federation/American Society of Civil Engineers) (1992) *Design of municipal wastewater treatment plants*.

WHO (World Health Organization) (1989) *Health Guidelines for the Use of Wastewater in Agriculture and Aquaculture*. Technical Report Series 778 (Geneva, 1989).

WPCF (Water Pollution Control Federation) (1991) Operation of municipal wastewater treatment plants. *MOP* 11 (III).

Lightning Source UK Ltd.
Milton Keynes UK
UKOW020328230212

187791UK00001B/52/A